Problem Solvers

Edited by L. Marder

Senior Lecturer in Mathematics, University of Southampton

No. 12

Fourier Series and Boundary-Value Problems

Problem Solvers

Fourier Series and Boundary-Value Problems

W. E. WILLIAMS

Professor of Mathematics
University of Surrey

LONDON · GEORGE ALLEN & UNWIN LTD

RUSKIN HOUSE   MUSEUM STREET

First published in 1973

ISBN 0 04 512022 6 *hardback*
0 04 512023 4 *paperback*

Printed in Great Britain
in 10 on 12 pt 'Monophoto' Times Mathematical Series 569
by Page Bros (Norwich) Ltd., Norwich

Contents

Acknowledgement

I should like to thank Dr. L. Marder for his helpful comments on early versions of the text and my colleague, Dr. J. H. Wilkinson, for his assistance in checking the proofs.

Chapter 1

Fourier Series

1.1 Piecewise Smooth Functions A function $f(x)$ is said to be piecewise continuous in the interval $[a, b]$ (the notation $[a, b]$ will be used to denote the closed interval $a \leqslant x \leqslant b$) if there exist a finite number of points $x_1, x_2, \ldots, x_n$ $(a = x_1 < x_2 \ldots < x_n = b)$ such that $f(x)$ is continuous in each sub-interval $x_i < x < x_{i+1}$ and has finite one-sided limits at the end points of each interval; these left- and right-hand limits will be denoted by $f(x_i + 0)$ and $f(x_{i+1} - 0)$ respectively. Clearly f need not necessarily even be defined at the end points of the sub-intervals.

A function f is said to be piecewise smooth in the interval $[a, b]$ if f and its first derivative f' are both piecewise continuous in this interval.

Problem 1.1 Show that, in the following, $f(x)$ is piecewise smooth in its interval of definition. Determine the points at which f or f' are discontinuous.

(a) $f(x) = x + c$, $(-c \leqslant x \leqslant 0)$, $f(x) = c - x$, $(0 \leqslant x \leqslant c)$.
(b) $f(x) = 0$, $(-1 \leqslant x \leqslant 0)$, $f(x) = x$, $(0 \leqslant x \leqslant 1)$.

Solution. (a) f is obviously continuous in each of the intervals $-c \leqslant x < 0$, $0 < x \leqslant c$ and $f(0 + 0) = f(0 - 0) = c$; thus f is continuous for x in $[-c, c]$. For $x < 0$, $f' = 1$ and for $x > 0$, $f' = -1$; hence f' is continuous in each of the sub-intervals $-c \leqslant x < 0, 0 < x \leqslant c$, but is discontinuous at $x = 0$. f' is not defined at the origin but both $f'(0 + 0)$ and $f'(0 - 0)$ exist and are finite (being ∓ 1 respectively). Thus f' is piecewise continuous and f is piecewise smooth in $[-c, c]$. The graph in Fig 1 represents f.

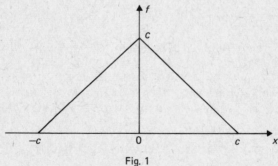

Fig. 1

1

(b) f is clearly continuous in each of the regions $-1 \leqslant x < 0$, $0 < x \leqslant 1$ and $f(0+0)$, $f(0-0)$ are both equal to zero. Also $f' = 0$ for $-1 \leqslant x < 0$ and $f' = 1$ for $0 < x \leqslant 1$, so that f' is discontinuous at $x = 0$. f' is not defined at $x = 0$ but both $f'(0-0)$ and $f'(0+0)$ exist and are finite, being equal to 0 and 1 respectively. Thus f is a piecewise smooth function. ☐

1.2 Fourier Series If a function $f(x)$ is integrable over the interval $[a, a+2c]$ then the Fourier series of f on the interval is the trigonometric series

$$\frac{1}{2} a_0 + \sum_{n=1}^{\infty} \left(a_n \cos \frac{n\pi x}{c} + b_n \sin \frac{n\pi x}{c} \right), \tag{1.1}$$

where
$$a_n = \frac{1}{c} \int_a^{a+2c} f(x) \cos \frac{n\pi x}{c} \, dx, \tag{1.2}$$

$$b_n = \frac{1}{c} \int_a^{a+2c} f(x) \sin \frac{n\pi x}{c} \, dx. \tag{1.3}$$

Fourier series are often defined over one of the intervals $[0, 2\pi]$, $[-\pi, \pi]$, the results for these are obtained from (1.2), (1.3) with appropriate choices of a and c. The a_r and b_r are known as the Fourier coefficients of f.

[An alternative form for the Fourier series can be obtained by expressing $\cos n\pi x/c$, $\sin n\pi x/c$, in terms of $\exp(\pm in\pi x/c)$ and this gives

$$\alpha_0 + \sum_{n=1}^{\infty} \alpha_n \exp(in\pi x/c) + \sum_{n=1}^{\infty} \bar{\alpha}_n \exp(-in\pi x/c),$$

where the bar denotes the complex conjugate and

$$\alpha_n = \frac{1}{2c} \int_a^{a+2c} f(x) \exp(-in\pi x/c) \, dx.$$

This form can be further simplified by replacing n by $-n$ in the second series, giving $\sum_{n=-\infty}^{\infty} \alpha_n \exp(in\pi x/c)$].

Theorem 1 If $f(x)$ is piecewise smooth in the interval $[a, a+2c]$ then the Fourier series of f on $[a, a+2c]$ converges for all x in the interval. The sum, $S(x)$, is equal to $f(x)$ at all points where f is continuous; if f has a discontinuity at $x = x_r$ then $2S(x_r) = f(x_r+0) + f(x_r-0)$. At the end points $2S$ takes the value $f(a+0) + f(a+2c-0)$.

Theorem 2 (*Parseval's theorem*). If f is piecewise continuous in $[a, a+2c]$ then

$$\frac{1}{c} \int_a^{a+2c} f^2 \, dx = \tfrac{1}{2} a_0^2 + \sum_{n=1}^\infty (a_n^2 + b_n^2).$$

Problem 1.2 (i) Find the Fourier series of $f(x) \equiv x$ on $[0, 2\pi]$ and determine the range of values of x for which the Fourier series converges to $f(x)$.

(ii) Sketch the function represented by the Fourier series in $[-2\pi, 4\pi]$.

Solution. (i) The coefficients a_n and b_n are found by setting $a = 0$ and $c = \pi$ in (1.2) and (1.3).

$$a_n = \frac{1}{\pi} \int_0^{2\pi} x \cos nx \, dx, \qquad b_n = \frac{1}{\pi} \int_0^{2\pi} x \sin nx \, dx,$$

$$a_0 = \frac{1}{\pi} \int_0^{2\pi} x \, dx = 2\pi.$$

The integrals can be evaluated by integration by parts.

$$a_n = \left[\frac{x}{\pi n} \sin nx \right]_{x=0}^{x=2\pi} - \frac{1}{n\pi} \int_0^{2\pi} \sin nx \, dx, \qquad n \neq 0,$$

$$= \frac{1}{n^2 \pi} \sin 2n\pi = 0,$$

$$b_n = -\left[\frac{x}{\pi n} \cos nx \right]_{x=0}^{x=2\pi} + \frac{1}{n\pi} \int_0^{2\pi} \cos nx \, dx,$$

$$= -\frac{2}{n} + \frac{1}{n^2 \pi} \sin 2n\pi = -\frac{2}{n}.$$

Hence setting $c = \pi$ in (1.1) shows that the Fourier series of x in $[0, 2\pi]$ is

$$\pi - 2 \sum_{n=1}^\infty \frac{\sin nx}{n}.$$

(ii) x is continuous and differentiable throughout the interval $[0, 2\pi]$ and hence, by Theorem 1, the Fourier series converges to x at all interior points. The value of the sum at the end points of the interval is the mean of the limiting values of x, and so the sum at each end point is π. The Fourier series has period 2π [i.e. $S(x) = S(x+2\pi)$] because each term has this period. Thus the sum $S(x)$ represents a function of period 2π, equal to x in $0 < x < 2\pi$ and to π at $x = 0$, 2π. S is shown in Fig. 2, where dots denote values at $x = -2\pi, 0, 2\pi, 4\pi$. □

Generally, if $S(x) = f(x)$ for $0 < x < 2c$, then, outside the interval, $S(x)$ will represent that function of period $2c$ which is equal to $f(x)$ on $0 < x < 2c$ and the Fourier series can be regarded as representing,

3

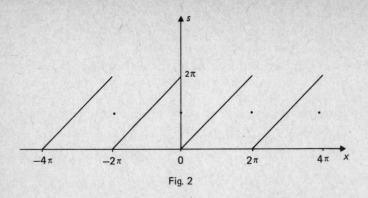

Fig. 2

outside the interval $0 < x < 2c$, the periodic extension of f. The Fourier series cannot represent a function $f(x)$ for all x unless f has period $2c$.

Problem 1.3 Find the Fourier series of $\cos 4\pi x/c$ in $[0, 2c]$.

Solution. In this case a in (1.2) and (1.3) is zero and hence

$$a_n = \frac{1}{c} \int_0^{2c} \cos \frac{n\pi x}{c} \cos \frac{4\pi x}{c} \, dx$$

$$= \frac{1}{2c} \int_0^{2c} \left[\cos \frac{(n-4)\pi x}{c} + \cos \frac{(n+4)\pi x}{c} \right] dx,$$

$$b_n = \frac{1}{c} \int_0^{2c} \sin \frac{n\pi x}{c} \cos \frac{4\pi x}{c} \, dx$$

$$= \frac{1}{2c} \int_0^{2c} \left[\sin \frac{(n-4)\pi x}{c} + \sin \frac{(n+4)\pi x}{c} \right] dx.$$

For $n \neq 4$ the integrals are all zero and hence $a_n = b_n = 0, n \neq 4$. Setting $n = 4$ gives $a_4 = 1$ and $b_4 = 0$, i.e. the Fourier series of $f(x) \equiv \cos 4\pi x/c$ is $\cos 4\pi x/c$. □

This result illustrates an important theorem: If two trigonometric series

$$\sum_{n=0}^{\infty} \left(a_n \cos \frac{n\pi x}{c} + b_n \sin \frac{n\pi x}{c} \right), \qquad \sum_{n=0}^{\infty} \left(a'_n \cos \frac{n\pi x}{c} + b'_n \sin \frac{n\pi x}{c} \right),$$

converge to the same sum at all but at most a finite number of points in $[a, a+2c]$ then $a_n = a'_n$, $b_n = b'_n$. An equivalent statement is that if a Fourier series is identically zero then so are all the coefficients.

Another corollary is that the Fourier series of any finite series of the form $\sum_{n=0}^{N} [a_n \cos (n\pi x/c) + b_n \sin (n\pi x/c)]$ is the series itself.

This latter result can be proved by direct calculation of the Fourier coefficients, the coefficient of $\cos m\pi x/c$ in the Fourier series is, by definition.

$$\frac{1}{c} \int_0^{2c} \left[\sum_{n=0}^{N} \left[a_n \cos\frac{n\pi x}{c} + b_n \sin\frac{n\pi x}{c} \right] \cos\frac{m\pi x}{c} \, dx. \right.$$

The following results can be verified to be valid:

$$\int_0^{2c} \cos\frac{n\pi x}{c} \cos\frac{m\pi x}{c} \, dx = \int_0^{2c} \sin\frac{n\pi x}{c} \cos\frac{m\pi x}{c} \, dx$$

$$= \int_0^{2c} \sin\frac{n\pi x}{c} \sin\frac{m\pi x}{c} \, dx = 0, \quad n \neq m,$$

$$\int_0^{2c} \cos^2\frac{n\pi x}{c} \, dx = \begin{cases} c, \ n \neq 0, \\ 2c, \ n = 0. \end{cases} \qquad \int_0^{2c} \sin^2\frac{n\pi x}{c} \, dx = \begin{cases} c, \ n \neq 0, \\ 0, \ n = 0. \end{cases}$$

Since the summation is over a finite number of terms the order of summation and integration can be interchanged and the above expressions show that the only non-zero contribution is that for $m = n$ giving that the Fourier coefficient of $\cos m\pi x/c$ is a_m ($m \leqslant N$) and zero otherwise.

A direct consequence of the above analysis is that if $f(x)$ is assumed to have an expansion of the form $\sum_{n=0}^{\infty} (a_n \cos(n\pi x/c) + b_n \sin(n\pi x/c))$ then carrying out the above analysis (formally replacing N by ∞) shows that a_n and b_n are the Fourier coefficients. This analysis is only formal as it has assumed a particular form for $f(x)$ and also assumed that integration and summation can be interchanged in an infinite series.

Problem 1.4 Find the Fourier series on $[-1, 1]$ of the function $f(x)$: $f(x) = 0, -1 \leqslant x \leqslant 0$; $f(x) = x, 0 \leqslant x \leqslant 1$. Hence show that

$$\sum_{r=1}^{\infty} (2r-1)^{-2} = \tfrac{1}{8}\pi^2.$$

Solution. In this example $a = -1$ and $c = 1$. The function $f(x)$ does not have the same form over the whole range in the integrals defining a_n and b_n. Thus the interval of integration has to be split into the two intervals $[-1, 0]$ and $[0, 1]$:

$$a_n = \int_{-1}^{1} f(x) \cos n\pi x \, dx = \int_0^1 x \cos n\pi x \, dx, \qquad [f \equiv 0, \ -1 \leqslant x \leqslant 0]$$

$$b_n = \int_{-1}^{1} f(x) \sin n\pi x \, dx = \int_0^1 x \sin n\pi x \, dx.$$

For $n \neq 0$ the integrals can be evaluated by integration by parts; the case $n = 0$ involves an elementary integral giving $a_0 = \frac{1}{2}$. Thus

$$a_n = \left[\frac{x \sin n\pi x}{n\pi}\right]_{x=0}^{x=1} - \frac{1}{n\pi}\int_0^1 \sin n\pi x\, dx = \frac{[\cos n\pi - 1]}{n^2\pi^2} = \frac{[(-1)^n - 1]}{n^2\pi^2};$$

$$b_n = -\left[\frac{x \cos n\pi x}{n\pi}\right]_{x=0}^{x=1} + \frac{1}{n\pi}\int_0^1 \cos n\pi x\, dx$$

$$= -\frac{\cos n\pi}{n\pi} + \frac{1}{n^2\pi^2}\sin n\pi = \frac{(-1)^{n+1}}{n\pi}.$$

Thus $a_n = 0$ (n even), and $a_{2r-1} = -2/(2r-1)^2\pi^2$, $r = 1, 2, \ldots$. The Fourier series is

$$\frac{1}{4} - \frac{2}{\pi^2}\sum_{r=1}^{\infty}\frac{\cos(2r-1)\pi x}{(2r-1)^2} + \sum_{n=1}^{\infty}\frac{(-1)^{n+1}\sin n\pi x}{n\pi}.$$

Since f is a piecewise smooth function [Problem 1.1(b)] it follows from Theorem 1 that the sum of the Fourier series at $x = 1$ is $\frac{1}{2}[f(1)+f(-1)] = \frac{1}{2}$. Substituting $x = 1$ in the Fourier series gives $\sum_{r=1}^{\infty}(2r-1)^{-2} = \frac{1}{8}\pi^2$. □

Problem 1.5 Find a Fourier series converging to $|(\sin x)|$ for every x.

Solution. A Fourier series cannot represent a function for all x unless it has the same period as the function (in this case π; see Problem 1.2). The precise interval is immaterial as both $|(\sin x)|$ and the relevant sines and cosines have the same period, and the integral of a periodic function over a cycle is independent of the end points. This can be proved as follows: Given $F(x) = F(x+2c)$ then (1–2), (1–3). *Hence*

$$\int_a^{a+2c} F(x)\, dx = \int_a^{2c} F(x)\, dx + \int_{2c}^{a+2c} F(x)\, dx =$$

$$= \int_a^{2c} F(x)\, dx + \int_{2c}^{a+2c} F(x+2c)\, dx$$

$$= \int_a^{2c} F(x)\, dx + \int_0^a F(x)\, dx = \int_0^{2c} F(x)\, dx.$$

Thus the integral is independent of a. In the present problem we choose the interval $[0, \pi]$ for simplicity, i.e. $a = 0$, $c = \frac{1}{2}\pi$ in (1.2), (1.3). Hence

$$a_n = \frac{2}{\pi}\int_0^{\pi}\sin x \cos 2nx\, dx = -\frac{4}{\pi(4n^2-1)}$$

$$b_n = \frac{2}{\pi}\int_0^{\pi}\sin x \sin 2nx\, dx = 0.$$

By Theorem 1 the Fourier series of $|\sin x|$ converges to $|\sin x|$ at all

6

interior points of $[0, \pi]$. At the end points the series converges to the mean of the end values of $|\sin x|$, i.e. to zero. It then follows by the periodicity of $|\sin x|$ that, for all x,

$$|\sin x| = \frac{2}{\pi} - \frac{4}{\pi} \sum_{n=1}^{\infty} \frac{\cos 2nx}{4n^2 - 1}. \qquad \square$$

Problem 1.6 If $f(x) = \sin x, 0 \leqslant x \leqslant \pi, f(x) = 0, \pi \leqslant x \leqslant 2\pi$, find the Fourier series of $f(x)$ on $[0, 2\pi]$.

Solution. This problem can be solved most easily by noticing that $f(x) = \frac{1}{2}(\sin x + |\sin x|)$. The Fourier series now follows directly from Problem 1.5 and the fact that $\sin x$ is its own Fourier series. The required Fourier series is

$$\tfrac{1}{2} \sin x + \frac{1}{\pi} - \frac{2}{\pi} \sum_{n=1}^{\infty} \frac{\cos 2nx}{(4n^2 - 1)}. \qquad \square$$

Problem 1.7 Find a Fourier series converging to $\sin^2 \frac{1}{2}x$ for all x.

Solution. $\sin^2 \frac{1}{2}x$ is a function of period 2π and, as in Problem 1.5, it suffices to determine its Fourier representation in $[0, 2\pi]$. This representation can be obtained most simply by writing $\sin^2 \frac{1}{2}x$ as $\frac{1}{2}(1 - \cos x)$ and noticing that, by the theorem quoted in Problem 1.3, the Fourier series of this function is itself. $\qquad \square$

1.3 Fourier Cosine and Sine Series In certain circumstances the Fourier series of a function $f(x)$ in $[-c, c]$ consists only of sine terms or only of cosine terms. This arises when

(i) $f(x)$ is an even function of x. [i.e. $f(x) = f(-x)$]—*cosine series*
(ii) $f(x)$ is an odd function of x. [i.e. $f(x) = -f(-x)$]—*sine series*.

This is illustrated in the following problem.

Problem 1.8 (i) Show that the Fourier series on $[-c, c]$ of an even function $f(x)$ contains no sine terms. (ii) Determine the general form of the Fourier series when $f(x)$ is an odd function.

Solution. (i) Setting $a = -c$ in (1.3) gives

$$cb_n = \int_{-c}^{c} f(x) \sin \frac{n\pi x}{c} \, dx.$$

The range of integration is now split into two ranges $[-c, 0]$ and $[0, c]$,

7

and x is replaced by $-x$ in the former. Thus

$$cb_n = \int_{-c}^{0} f(x) \sin \frac{n\pi x}{c} \, dx + \int_{0}^{c} f(x) \sin \frac{n\pi x}{c} \, dx$$

$$= -\int_{0}^{c} f(-x) \sin \frac{n\pi x}{c} \, dx + \int_{0}^{c} f(x) \sin \frac{n\pi x}{c} \, dx = 0,$$

since $f(x) = f(-x)$. Therefore the Fourier series will not contain any sine terms. From (1.1) we note also that

$$ca_n = \int_{-c}^{0} f(x) \cos \frac{n\pi x}{c} \, dx + \int_{0}^{c} f(x) \cos \frac{n\pi x}{c} \, dx$$

$$= \int_{0}^{c} f(-x) \cos \frac{n\pi x}{c} \, dx + \int_{0}^{c} f(x) \cos \frac{n\pi x}{c} \, dx = 2 \int_{0}^{c} f(x) \cos \frac{n\pi x}{c} \, dx.$$

(ii) For the case when f is odd, it follows from the above expressions for a_n and b_n, by replacing $f(-x)$ by $-f(x)$ that $a_n = 0$, $cb_n = 2 \int_{0}^{c} f(x) \sin (n\pi x/c) \, dx$, whence the result follows. □

Problem 1.9 Show that, in $[-\pi, \pi]$,

$$x^2 = \tfrac{1}{3}\pi^2 + 4 \sum_{n=1}^{\infty} \frac{(-1)^n \cos nx}{n^2}$$

and hence evaluate (i) $\sum_{n=1}^{\infty} n^{-2}$, (ii) $\sum_{n=1}^{\infty} (-1)^{n+1} n^{-2}$, (iii) $\sum_{n=1}^{\infty} n^{-4}$.

Solution. The first step is to determine the Fourier series of x^2 on $[-\pi, \pi]$. Since x^2 is even the series will consist only of cosine terms, and setting $c = \pi$ in the appropriate equation of Problem 1.8 shows that the Fourier series is

$$\tfrac{1}{2}a_0 + \sum_{n=1}^{\infty} a_n \cos nx \qquad \text{where} \qquad a_n = \frac{2}{\pi} \int_{0}^{\pi} x^2 \cos nx \, dx.$$

The integral can be evaluated by integration by parts for $n \neq 0$, giving $n^2 a_n = 4(-1)^n$; the integration for $n = 0$ is trivial and gives $3a_0 = 2\pi^2$. Since x^2 satisfies all the conditions of Theorem 1 and takes the same value at the end points of the interval $[-\pi, \pi]$, the sum of the Fourier series is equal to x^2 at all points in $[-\pi, \pi]$, i.e.

$$x^2 = \tfrac{1}{3}\pi^2 + 4 \sum_{n=1}^{\infty} [(-1)^n \cos nx/n^2], \qquad |x| \leqslant \pi.$$

(i) Setting $x = \pi$ gives, since $\cos n\pi = (-1)^n$, $\sum_{n=1}^{\infty} n^{-2} = \tfrac{1}{6}\pi^2$.

(ii) Similarly setting $x = 0$ gives $\sum\limits_{n=1}^{\infty} (-1)^{n+1} n^{-2} = \frac{1}{12}\pi^2$.

(iii) The determination of $\sum\limits_{n=1}^{\infty} n^{-4}$ is less straightforward, but can be made by noticing that the terms are squares of the coefficients of the Fourier series for x^2. This suggests use of Theorem 2 (Parseval's theorem) which gives

$$\frac{1}{\pi} \int_{-\pi}^{\pi} x^4 \, dx = \frac{\pi^4}{18} + 16 \sum_{n=1}^{\infty} n^{-4},$$

Hence
$$90 \sum_{n=1}^{\infty} n^{-4} = \pi^4. \qquad \square$$

For any function $f(x)$ defined in $[0, c]$ one can construct an even function $h(x)$ equal to $f(x)$ for $0 < x < c$ by setting $h(x) = f(-x)$ for $-c < x < 0$ and thus, by Problem 1.8, there will exist a Fourier series representation of $f(x)$ in $0 < x < c$ consisting only of cosine terms. Similarly by defining $h(x)$ by $-f(-x)$ for $-c < x < 0$ it will be possible to construct a Fourier series representation of $f(x)$ in $0 < x < c$ consisting only of sines. These representations are formally defined as follows:

Fourier cosine series If $f(x)$ is integrable on $\lceil 0, c \rceil$ then the Fourier cosine series of f on $[0, c]$ is defined to be

$$\frac{1}{2} a_0 + \sum_{n=1}^{\infty} a_n \cos \frac{n\pi x}{c},$$

where
$$a_n = \frac{2}{c} \int_0^c f(x) \cos \frac{n\pi x}{c} \, dx. \tag{1.4}$$

Fourier sine series If $f(x)$ is integrable on $[0, c]$ then the Fourier sine series of f on $[0, c]$ is defined to be

$$\sum_{n=1}^{\infty} b_n \sin \frac{n\pi x}{c},$$

where
$$b_n = \frac{2}{c} \int_0^c f(x) \sin \frac{n\pi x}{c} \, dx. \tag{1.5}$$

Theorem 3 The Fourier cosine and sine series of a function $f(x)$ which is piecewise smooth in $[0, c]$ converge to $f(x)$ at any point x in $0 < x < c$ where f is continuous. At any interior point $x = x_r$ where f is discontinuous, both series converge to $\frac{1}{2}[f(x_r + 0) + f(x_r - 0)]$. The cosine series converges to $f(0+0)$ at $x = 0$, to $f(c-0)$ at $x = c$ and is always continuous. The sine series converges to zero for $x = 0, c$. Theorem 3 can be

9

proved by applying the results of Problem 1.8 to the even and odd functions equal to $f(x)$ in $0 < x < c$. (Since these series only represent $f(x)$ on an interval whose length is half the period of the functions involved, the series are often referred to as half range series.)

It should be noted that a corollary of Theorem 3 is that for a function $f(x)$ continuous in $[0, c]$, the Fourier cosine series converges to f in this interval but the convergence of the sine series in the closed interval requires the additional conditions $f(0) = f(c) = 0$. Another corollary is that a function continuous in $[0, c]$ and vanishing at $x = 0$ and $x = c$ can be represented throughout the interval by its Fourier sine series.

Problem 1.10   Find, on the interval $[0, \pi]$, (a) the Fourier cosine series of x and (b) the Fourier sine series of $\cos \frac{1}{2}x$. Sketch the functions defined by the Fourier series for $|x| \leqslant 3\pi$.

Solution. (a) The Fourier cosine series of x is $\frac{1}{2}a_0 + \sum\limits_{n=1}^{\infty} a_n \cos nx$ where $a_n = (2/\pi) \int_0^\pi x \cos nx \, dx$. We find $a_0 = \pi$, for $n \neq 0$ integration by parts gives $n^2 a_n = 2[(-1)^n - 1]/\pi$. Thus a_n will vanish for even values of n. For $n = 2r - 1$ $(r = 1, 2, \ldots)$ we get $\pi(2r-1)^2 a_{2r-1} = -4$. The Fourier cosine series is thus

$$\tfrac{1}{2}\pi - \frac{4}{\pi} \sum_{r=1}^{\infty} \frac{\cos(2r-1)x}{(2r-1)^2}$$

and by Theorem 3 converges to x in $[0, \pi]$.

The cosine series is an even function of x and is equal to x in $[0, \pi]$ and to $-x$ in $[-\pi, 0]$. This even continuation (denoted by heavy lines) of x is shown in Fig. 3 together with its periodic extension (period 2π) to $[-3\pi, 3\pi]$.

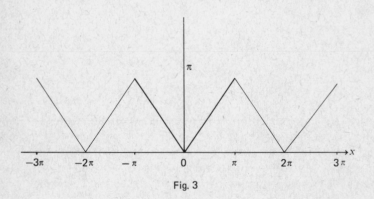

Fig. 3

(b) The coefficient b_n in the sine series of $\cos\frac{1}{2}x$ is

$$b_n = \frac{2}{\pi}\int_0^\pi \sin nx \cos\tfrac{1}{2}x\,dx = \frac{1}{\pi}\int_0^\pi \left[\sin(n+\tfrac{1}{2})x + \sin(n-\tfrac{1}{2})x\right]dx$$

$$= \frac{8n}{\pi(4n^2-1)}.$$

The series converges to $\cos\frac{1}{2}x$ at all interior points of $[0,\pi]$ and to zero at the end points and hence represents $\cos\frac{1}{2}x$ in $0 < x \leqslant \pi$.

The series defines an odd function of x and hence in $-\pi \leqslant x < 0$ is equal to $-\cos\frac{1}{2}x$. From this and the periodicity property it follows that the graph is as shown in Fig. 4 (the form of the series in $[-\pi, \pi]$ being denoted by heavy lines). □

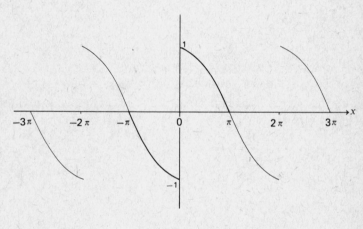

Fig. 4

Problem 1.11 (i) Determine the general form of the Fourier cosine and sine series in $[0, c]$ of a function $f(x)$ satisfying the relationship $f(c-x) = f(x)$.

(ii) Obtain the corresponding series when $f(c-x) = -f(x)$.

Solution. (i) We consider first the integral from $x = 0$ to $x = c$ of an arbitrary function $F(x)$ such that $F(c-x) = F(x)$. Such a function is said to be even about $\frac{1}{2}c$ as setting $x = \frac{1}{2}c - \varepsilon$ gives $F(\frac{1}{2}c+\varepsilon) = F(\frac{1}{2}c-\varepsilon)$, i.e. it takes the same value at a given distance on either side of $x = \frac{1}{2}c$. [If $F(c-x) = -F(x)$ then the function is said to be odd about $x = \frac{1}{2}c$.]

11

The range of integration is split into the two intervals $[0, \frac{1}{2}c], [\frac{1}{2}c, c]$, thus

$$\int_0^c F(x)\,dx = \int_0^{\frac{1}{2}c} F(x)\,dx + \int_{\frac{1}{2}c}^c F(x)\,dx$$
$$= \int_0^{\frac{1}{2}c} F(x)\,dx + \int_0^{\frac{1}{2}c} F(c-x)\,dx = 2\int_0^{\frac{1}{2}c} F(x)\,dx.$$

For a function $G(x)$ odd about $x = \frac{1}{2}c$ it follows that $\int_0^c G(x)\,dx = 0$. Sin $n\pi x/c$ is odd or even about $\frac{1}{2}c$ according as to whether n is even or odd and $\cos n\pi x/c$ is odd or even about $\frac{1}{2}c$ for according as to whether n is odd or even. Thus, for the given function $f(x)$, $f(x)\cos n\pi x/c$ is even or odd about $\frac{1}{2}c$ according as to whether n is even or odd and from the above general result it follows that $a_n = 0$ in the cosine series for n odd and

$$a_{2r} = \frac{4}{c}\int_0^{\frac{1}{2}c} f(x)\cos\frac{2\pi rx}{c}\,dx, \qquad r = 0, 1, \ldots.$$

$f(x)\sin n\pi x/c$ is odd or even about $\frac{1}{2}c$ according as to whether n is even or odd and thus $b_n = 0$ for n even and

$$b_{2r-1} = \frac{4}{c}\int_0^{\frac{1}{2}c} f(x)\sin(2r-1)\frac{\pi x}{c}\,dx, \qquad r = 1, 2, \ldots.$$

(ii) In this case $f(x)\cos n\pi x/c$ is even or odd according as to whether n is odd or even and thus $a_n = 0$ for n even and

$$a_{2r-1} = \frac{4}{c}\int_0^{\frac{1}{2}c} f(x)\cos(2r-1)\frac{\pi x}{c}\,dx, \qquad r = 1, 2, \ldots.$$

Similarly $b_n = 0$ for n odd and

$$b_{2r} = \frac{4}{c}\int_0^{\frac{1}{2}c} f(x)\sin\frac{2r\pi x}{c}\,dx, \qquad r = 1, 2, \ldots. \qquad \square$$

Problem 1.12 Find the sine series for $\cos x$ in $[0, \pi]$.

Solution. Cos x is odd about $\frac{1}{2}\pi$ and hence, from Problem 1.11, the sine series will be of the form $\sum_{r=1}^{\infty} b_{2r}\sin 2rx$ where

$$b_{2r} = \frac{4}{\pi}\int_0^{\frac{1}{2}\pi}\cos x\sin 2rx\,dx = \frac{8r}{\pi(4r^2-1)}.$$

The sine series converges to zero at $x = 0$ and $x = \pi$ and hence

$$\cos x = \frac{8}{\pi}\sum_{r=1}^{\infty}\frac{r\sin 2rx}{4r^2-1}, \qquad 0 < x < \pi. \qquad \square$$

Replacing x by $\tfrac{1}{2}x$ gives

$$\cos \tfrac{1}{2}x = \frac{8}{\pi} \sum_{r=1}^{\infty} \frac{r\sin rx}{4r^2-1}, \qquad 0 < x < 2\pi.$$

This latter result is in agreement with one previously obtained in Problem 1.10.

Problem 1.13 Find the sine and cosine series on $[0, c]$ of the function $f(x): f(x) = x, 0 \leqslant x \leqslant \tfrac{1}{2}c; f(x) = c-x, \tfrac{1}{2}c \leqslant x \leqslant c$, and hence evaluate $\sum_{r=1}^{\infty} (2r-1)^{-2}$. Sketch the functions defined by the Fourier series in $|x| \leqslant 2c$.

Solution. The cosine series is $\tfrac{1}{2}a_0 + \sum_{n=1}^{\infty} a_n \cos(n\pi x/c)$ where

$$a_n = \frac{2}{c} \int_0^c f(x) \cos \frac{n\pi x}{c} \, dx.$$

$f(x)$ does not have the same form over the whole interval of integration and this interval has to be split into the two intervals $[0, \tfrac{1}{2}c]$, $[\tfrac{1}{2}c, c]$. f is even about $x = \tfrac{1}{2}c$ and from Problem 1.11 it follows that $a_n = 0$ for odd values of n. However in order to illustrate the point that the correct Fourier series can be calculated without direct use of the evenness property the integral for a_n will be evaluated directly.

$$\frac{ca_n}{2} = \int_0^{\frac{1}{2}c} x \cos \frac{n\pi x}{c} \, dx + \int_{\frac{1}{2}c}^{c} (c-x)\cos \frac{n\pi x}{c} \, dx, \qquad n \neq 0$$

$$= \frac{c^2}{n^2\pi^2} \left[2\cos \frac{n\pi}{2} - 1 - (-1)^n \right], \qquad n \neq 0 \text{ (by integration by parts)},$$

$a_0 = \tfrac{1}{2}c$.

Clearly a_n vanishes for n odd and hence

$$a_{2r} = \frac{c}{\pi^2 r^2} (\cos r\pi - 1).$$

This vanishes for r even and the only non-zero values are for $r = 2s-1$ $[s = 1, 2, \dots]$ when $a_{4s-2} = -2c/\pi^2(2s-1)^2$, the Fourier series is thus

$$\frac{1}{4}c - \frac{2c}{\pi^2} \sum_{s=1}^{\infty} \frac{\cos(4s-2)\pi x/c}{(2s-1)^2}.$$

$f(x)$ is piecewise smooth throughout $[0, c]$. Hence, by Theorem 3, the Fourier series converges to f on $[0, c]$ and setting $x = 0$ gives

$$\sum_{s=1}^{\infty} (2s-1)^{-2} = \tfrac{1}{8}\pi^2.$$

13

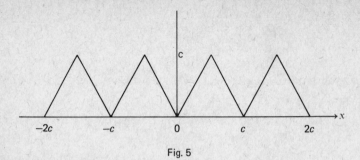

Fig. 5

The cosine series represents an even function of x equal to $f(x)$ in $[0, c]$ and its form is shown in Fig. 5. For $c = 2\pi$ this has the same form as the Fourier series of Problem 1.10 (a).

The coefficients b_n can also be calculated directly as above or by using the evenness property of $f(x)$. Problem 1.11 shows that $b_n = 0$ (n even) and

$$cb_{2r-1} = 4 \int_0^{\frac{1}{2}c} x \sin(2r-1) \frac{\pi x}{c} dx, \quad r = 1, 2, \ldots.$$

$$= \frac{4(-1)^{r-1}c^2}{(2r-1)^2\pi^2}, \qquad \text{[by integration by parts]}.$$

$f(x)$ vanishes at $x = 0$ and $x = c$ and thus, in $[0, c]$,

$$f(x) = \frac{4c}{\pi^2} \sum_{r=1}^{\infty} \frac{(-1)^{r-1}}{(2r-1)^2} \sin(2r-1)\frac{\pi x}{c}.$$

The sine series represents an odd function of x equal to $f(x)$ on $[0, c]$ and of period $2c$; its form is shown in Fig. 6.

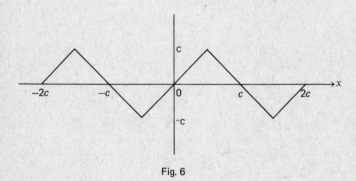

Fig. 6

1.4 Cosine and Sine Series of Period 4c In applications it is often necessary to represent a function $f(x)$ on $[0, c]$ as a cosine or sine series whose general terms are $a_{2n-1}\cos(2n-1)\pi x/2c$ and $b_{2n-1}\sin(2n-1)\pi x/2c$ respectively. From Problem 1.11 it is seen that such representations could be obtained on $[0, 2c]$ for functions odd or even about $x = c$, and thus all that it is necessary to do is to extend the given function to $[0, 2c]$ and make it even or odd about $x = c$ as necessary. The coefficients a_{2n-1}, b_{2n-1} can be obtained by replacing c by $2c$ in Problem 1.11, giving

$$a_{2n-1} = \frac{2}{c}\int_0^c f(x)\cos\frac{(2n-1)\pi x}{2c}\,dx,$$

$$b_{2n-1} = \frac{2}{c}\int_0^c f(x)\sin\frac{(2n-1)\pi x}{2c}\,dx.$$

The even extension of f is continuous and thus the sine series converges to f in $0 < x \leqslant c$ and to zero at $x = 0$; the odd extension is discontinuous at $x = c$ and the cosine series converges to f in $0 \leqslant x < c$ and to zero at $x = c$.

Problem 1.14 Determine the Fourier cosine series and sine series of period 4π which represent x in $[0, \pi]$ and sketch the functions represented by the series in $|x| \leqslant 4\pi$.

Solution. Setting $c = \pi$ in the preceding analysis gives

$$x = \sum_{n=1}^{\infty} a_{2n-1}\cos\tfrac{1}{2}(2n-1)x, \qquad 0 \leqslant x < \pi,$$

$$x = \sum_{n=1}^{\infty} b_{2n-1}\sin\tfrac{1}{2}(2n-1)x, \qquad 0 < x \leqslant \pi,$$

where

$$a_{2n-1} = \frac{2}{\pi}\int_0^\pi x\cos\tfrac{1}{2}(2n-1)x\,dx, \quad b_{2n-1} = \frac{2}{\pi}\int_0^\pi x\sin\tfrac{1}{2}(2n-1)x\,dx.$$

The integrals can be very easily evaluated by integration by parts giving

$$a_{2n-1} = \frac{4(-1)^{n+1}}{(2n-1)} - \frac{8}{\pi(2n-1)^2}, \qquad b_{2n-1} = \frac{8(-1)^{n+1}}{\pi(2n-1)^2}.$$

x vanishes at $x = 0$ and hence the sine series converges to x in $[0, \pi]$. $\cos\tfrac{1}{2}(2n-1)(\pi-x) = -\cos\tfrac{1}{2}(2n-1)(\pi+x)$ and thus the cosine series represents a function odd about $x = \pi$ (this of course also follows from the discussion preceding this example). Thus the form of the cosine series is known for $0 \leqslant x \leqslant 2\pi$ and, as the cosine series is an even function of x, it can be now determined for $-2\pi \leqslant x \leqslant 0$. The series is thus known for a complete cycle and the form for all x can now be obtained and is

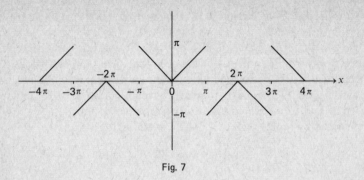

Fig. 7

given in Fig. 7. The form for the sine series is obtained similarly and is given in Fig. 8. ☐

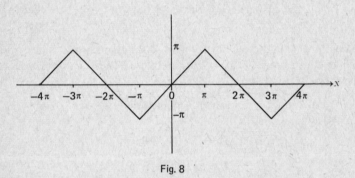

Fig. 8

These results provide two more examples of Fourier series expansions which represent the same function over a given interval but which represent very different functions when considered for all x (cf. Problem 1.13).

Problem 1.15 Find the representation in $[0, c]$ of $\sin \pi x / c$ as a Fourier cosine series of period $4c$.

Solution. The coefficient a_{2n-1} of $\cos (2n-1)\pi x/c$ in the expansion is given by

$$a_{2n-1} = \frac{2}{c} \int_0^c \sin \frac{\pi x}{c} \cos \frac{1}{2}(2n-1)\frac{\pi x}{c} \, dx.$$

The integral can be evaluated by expressing the integrand as a sum of sines and this gives $a_n = -8/\pi(2n+1)(2n-3)$. ☐

16

1.5 Differentiation and Integration of Fourier Series Differentiation of a Fourier series will produce another Fourier series whose coefficients are proportional to na_n, nb_n where a_n and b_n are the coefficients of the original series. Thus, even if the original series converges, the differentiated series may not even converge if a_n and b_n are not sufficiently small for large n and it is therefore useful to be able to derive, from the form of the original function, the behaviour of the coefficients for large n and we have the following theorems:

Theorem 4 If f is piecewise smooth in $[a, a+2c]$ then a_n and b_n are at most $O(n^{-1})$ (i.e. $\lim_{n \to \infty} na_n$, $\lim_{n \to \infty} nb_n$ are both finite).

Theorem 5 If f and its first $(p-1)$ derivatives are continuous in $[a, a+2c]$; $f^{(r)}(a) = f^{(r)}(a+2c)$, $r = 0, 1, \ldots p-1$, (where $f^{(r)}$ is the rth derivative of f) and $f^{(p)}$ is piecewise smooth in $[a, a+2c]$ then a_n and b_n are at most $O(n^{-p-1})$.

Problem 1.16 Show that, for large n, the coefficients of the cosine series on $[0, \pi]$ of a function $f(x)$, continuous with a piecewise smooth first derivative, are $O(n^{-2})$.

Solution. The behaviour of the coefficients cannot be deduced directly from the above theorems which refer to the complete Fourier expansion of a function. The cosine series expansion of f is the complete Fourier expansion on $[-\pi, \pi]$ of the extended function even in x and equal to f in $[0, \pi]$. This function is continuous in $[-\pi, \pi]$, taking the same value at both end points, and hence Theorem 5 may be applied with $p = 1$, giving that the Fourier coefficients are $O(n^{-2})$. ☐

The derivative of the extended function is not continuous at $x = 0$ as its left and right hand limits are $\pm f'(0)$ respectively, at the end points this derivative is $\pm f'(\pi)$. Thus, if $f'(0) = f'(\pi) = 0$, Theorem 5 shows that the Fourier coefficients of the cosine series for f would be $O(n^{-3})$ if f' were continuous and f'' piecewise smooth.

Problems 1.10(a), 1.13, 1.15 are all examples on deriving cosine series of functions satisfying the conditions of the present example and all have Fourier coefficients of $O(n^{-2})$. It follows that the coefficients of the sine series obtained by differentiating cosine series of the above function are $O(n^{-1})$ and the series is at least convergent. This however does not prove that the differentiated series converges to $f'(x)$. Similarly the coefficients of the Fourier sine series of f are $O(n^{-1})$ but will be $O(n^{-2})$ provided that $f(0) = f(\pi) = 0$. Thus in many cases the series obtained on differentiating the sine series of a continuous function will not even converge.

17

The general conditions concerning differentiation of a Fourier series are contained in the following theorem:

Theorem 6   If $f(x)$ is continuous and $f'(x)$ piecewise smooth in $[0, c]$ then
  (i) The Fourier cosine series for f may be differentiated term by term and the resulting series converges to f' in $0 < x < c$.
  (ii) The Fourier sine series may be differentiated term by term if, and only if, $f(0) = f(c) = 0$, and the differentiated series then converges to f' in $[0, c]$.
For the case when $f(0)$ or $f(c)$ are not zero the coefficients of the cosine series for f' are given by

$$a_0 = \frac{2}{c}[f(c) - f(0)], \qquad a_n = \frac{n\pi b_n}{c} - \frac{2}{c}[f(0) + (-1)^{n+1} f(c)].$$

In applying this theorem to cases where the derivative of f is not defined at a point $x = x_0$, f' at that point is to be interpreted as $\frac{1}{2}[f'(x_0+0) + f'(x_0-0)]$. The theorem is proved by using integration by parts to express the Fourier coefficients of f' in terms of those for f.

Problem 1.17   Use the results of Problem 1.10(b) to obtain the Fourier cosine series of $\sin \frac{1}{2}x$ on $[0, \pi]$.

Solution. Problem 1.10(b) shows that

$$\cos\frac{1}{2}x = \frac{8}{\pi} \sum_{n=1}^{\infty} \frac{n \sin nx}{4n^2 - 1}, \qquad 0 < x \leqslant \pi.$$

The Fourier coefficients a_n of $-\frac{1}{2}\sin\frac{1}{2}x$ are, from Theorem 6(ii),

$$a_0 = -\frac{2}{\pi}, \qquad a_n = \frac{8n^2}{\pi(4n^2 - 1)} - \frac{2}{\pi} = \frac{2}{\pi(4n^2 - 1)}.$$

Thus

$$\sin\frac{1}{2}x = \frac{2}{\pi} - \frac{4}{\pi} \sum_{n=1}^{\infty} \frac{\cos nx}{(4n^2 - 1)}, \qquad 0 < x \leqslant \pi. \qquad \square$$

Problem 1.18   Use the results of Problem 1.13 to determine the cosine and sine series in $[0, c]$ of the function $g(x)$: $g(x) = 1, 0 \leqslant x < \frac{1}{2}c$; $g(\frac{1}{2}c) = 0$; $g(x) = -1, \frac{1}{2}c < x \leqslant c$.

Solution. Except at $x = \frac{1}{2}c$ the function g is clearly the derivative of f defined in Problem 1.13. At $x = \frac{1}{2}c$, however, f' is not defined and thus f' at $x = \frac{1}{2}c$, for application of Theorem 6, has to be interpreted as the mean of the limiting values of the derivative on either side of $x = \frac{1}{2}c$, i.e. zero. Theorem 6(i) then shows that the sine series for g is $4\sum_{s=1}^{\infty} [\sin(4s - 2)\pi x/c]/(2s - 1)$. Theorem 6(ii) shows that the coefficients

in the cosine series of f' are given by $a_0 = 0[f(0) = f(c) = 0], a_n = n\pi b_n/c,$
where
$$b_n = 0 \quad (n \text{ even}), \qquad b_{2r-1} = \frac{4(-1)^{r-1}c}{(2r-1)^2\pi^2}, \quad r = 1, 2, \ldots.$$
This gives $a_n = 0$ (n even) and $a_{2r-1} = 4(-1)^{r-1}/(2r-1)\pi, r = 1, 2 \ldots$ □

Results analogous to those of Theorem 6 can be obtained for the cosine and sine series of period $4c$ defined in section 1.3.

Theorem 7 If f is continuous in $[0, c]$ with a continuous first derivative then

  (i) when f is represented in $[0, c]$ by the series

$$\sum_{r=1}^{\infty} a_{2r-1} \cos(2r-1)\frac{\pi x}{2c}$$

the derivative f' can be represented by the series

$$\sum_{r=1}^{\infty} b_{2r-1} \sin(2r-1)\frac{\pi x}{2c}$$

where
$$b_{2r-1} = \frac{2(-1)^{r+1}}{c} f(c) - \frac{(2r-1)}{2c}\pi a_{2r-1}.$$

  (ii) when f is represented in $[0, c]$ by the series

$$\sum_{r=1}^{\infty} b_{2r-1} \sin(2r-1)\frac{\pi x}{2c}$$

the derivative f' can be represented by the series

$$\sum_{r=1}^{\infty} a_{2r-1} \cos(2r-1)\frac{\pi x}{2c}$$

where
$$a_{2r-1} = -\frac{2}{c} f(0) + \frac{(2r-1)\pi b_{2r-1}}{2c}.$$

This theorem is also proved by means of integration by parts.

Integration of Fourier series The situation concerning the integration of Fourier series is much simpler than that for differentiation and it can be proved that a Fourier series of a function f may be integrated term by term and that the integrated series converges to the integral of f.

Problem 1.19 Show that the Fourier cosine series of $\sin\frac{1}{2}x$ on $[0, \pi]$ may be obtained by integration from Problem 1.10.

  Solution. In Problem 1.10 it is shown that

$$\cos\frac{1}{2}t = \frac{8}{\pi} \sum_{n=1}^{\infty} \frac{n}{4n^2-1} \sin nt, \quad 0 < t \leqslant \pi,$$

and integrating this identity from $t = 0$ to $t = x$ gives

$$2 \sin \tfrac{1}{2}x = -\frac{8}{\pi} \sum_{n=1}^{\infty} \frac{(\cos nx - 1)}{4n^2 - 1}, \qquad 0 \leqslant x \leqslant \pi.$$

It is known that $\sum_{n=1}^{\infty} (4n^2 - 1)^{-1} = \tfrac{1}{2}$ and thus

$$\sin \tfrac{1}{2}x = \frac{2}{\pi} - \frac{4}{\pi} \sum_{n=1}^{\infty} \frac{\cos nx}{4n^2 - 1}.$$

$$\left[\text{The sum } \sum_{n=1}^{\infty} (4n^2 - 1)^{-1} \text{ can be calculated by setting } x = 0 \text{ in the co-} \right.$$
sine series for $\sin x$ in $[0, \pi]$:

$$\left. \sin x = \frac{2}{\pi} - \frac{4}{\pi} \sum_{n=1}^{\infty} \frac{\cos 2nx}{4n^2 - 1}, \qquad 0 \leqslant x \leqslant \pi. \right]$$ $\qquad \square$

For $a_0 \neq 0$ the integrated series will not be a Fourier series.

EXERCISES

1. Show that

(i) $x^2 = \dfrac{4\pi^2}{3} + 4 \displaystyle\sum_{n=1}^{\infty} \dfrac{\cos nx}{n^2} - 4\pi \displaystyle\sum_{n=1}^{\infty} \dfrac{\sin nx}{n}, \quad 0 < x < 2\pi.$

(ii) $\exp ax = \dfrac{2 \sinh a\pi}{\pi} \left[\dfrac{1}{2a} + \displaystyle\sum_{n=1}^{\infty} \dfrac{(-1)^n a \cos nx - n \sin nx}{a^2 + n^2} \right],$

$-\pi < x < \pi.$

By considering the value of the series at the end points show that

$$\tfrac{1}{2} \pi a \coth \pi a = \frac{1}{2} + \sum_{n=1}^{\infty} \frac{a^2}{(a^2 + n^2)}.$$

(iii) $\tfrac{1}{12} x(\pi - x)(2\pi - x) = \displaystyle\sum_{n=1}^{\infty} \dfrac{\sin nx}{n^3}, \quad 0 \leqslant x \leqslant 2\pi.$

(iv) $|x| = \tfrac{1}{2}\pi - \dfrac{4}{\pi} \displaystyle\sum_{n=1}^{\infty} \dfrac{\cos(2n-1)x}{(2n-1)^2}, \quad -\pi \leqslant x \leqslant \pi,$

and hence evaluate $\displaystyle\sum_{n=1}^{\infty} (2n-1)^{-2}, \displaystyle\sum_{n=1}^{\infty} (2n-1)^{-4}.$

2. Determine on the interval $[0, \pi]$ the sine series for x and the 'cosine' series for $\cos \tfrac{1}{2}x$ and obtain the range of values of x for which the series converge to the relevant functions.

3. Obtain by direct calculation the Fourier cosine and sine series in $[0, c]$ of the function $g(x)$ defined in Problem 1.18.

4. Find the cosine and sine series of period $4c$ representing $\exp x$ in $0 < x < c$.

5. Determine the cosine series in $[0, \pi]$ for $\sin x$ by applying Theorem 6 to the sine series for $\cos x$ derived in Problem 1.12.

Chapter 2

Solution of Boundary-Value Problems by Means of Fourier Series

2.1 Introduction Most of the subsequent problems will be concerned with the solution of the simpler partial differential equations of mathematical physics. Particular examples of these equations are:

$\nabla^2 u = \rho$ (Poisson's equation), $\nabla^2 u = 0$ (Laplace's equation),

$\nabla^2 u = \dfrac{1}{c^2}\dfrac{\partial^2 u}{\partial t^2}$ (wave equation), $\nabla^2 u = k\dfrac{\partial u}{\partial t}$ (diffusion equation),

$\nabla^2 u = ku$ (Helmholtz' equation).

For two independent variables (which is the case with which we shall mainly be concerned) all the above are particular cases of

$$a_1(x)\frac{\partial^2 u}{\partial x^2}+a_2(x)\frac{\partial u}{\partial x}+a_3(x)u+b_1(y)\frac{\partial^2 u}{\partial y^2}+b_2(y)\frac{\partial u}{\partial y}+b_3(y)u = f(x,y), \quad (2.1)$$

where a_i, b_i are known functions of x and y respectively and f is another known function of these variables.

The methods to be described are only suitable for solving (2.1) in a rectangular domain D of the x, y plane and in the present chapter it is necessary that the range of one independent variable is finite and that the coefficient of the second derivative with respect to that variable does not vanish within D. Without loss of generality x is taken to be that variable (i.e. $a_1 \neq 0$ in D) and D to be the domain $0 \leqslant x \leqslant c$, $b \leqslant y \leqslant d$; either or both of b, d can be unbounded.

The boundary conditions under which (2.1) is soluble depend critically on the sign of b_1/a_1 (it will be assumed this sign does not change within D). [An equivalent method of relating the boundary conditions to the equation is by reference to the classification of partial differential equations in two independent variables (cf. P. M. Morse and H. Feshbach, *Methods of Theoretical Physics*, McGraw-Hill, 1953, Ch. 6). In this nomenclature, b_1/a_1 being positive, negative or zero determines whether (2.1) is elliptic, hyperbolic or parabolic.]

The boundary value problems to be solved in subsequent problems will be such that they possess unique solutions and the following table will, for most practical situations, indicate the appropriate conditions to be applied.

22

$b_1/a_1,$ + ve u or $\partial u/\partial y$ given on $y = b, y = d.$

$b_1/a_1,$ − ve $u, \partial u/\partial y$ given on $y = b.$

$b_1/a_1, 0$ u given on $y = b.$

$\left.\begin{array}{c} \\ \\ \\ \end{array}\right\}$ u or $\partial u/\partial x$ given for $x = 0, c.$

2.2 Homogeneous Problems

Homogeneous problems in the present context are defined to be those of solving

$$a_1 \frac{\partial^2 u}{\partial x^2} + a_2 \frac{\partial u}{\partial x} + a_3 u + b_1 \frac{\partial^2 u}{\partial y^2} + b_2 \frac{\partial u}{\partial y} + b_3 u = 0, \qquad (2.2)$$

in D with u or $\partial u/\partial x$ vanishing on $x = 0, c$. The use of Fourier series (and their generalizations) to solve (2.2) relies on two separate concepts. The first is the principle of superposition which states that if any number of functions separately satisfy (2.2) then so will any linear combination of these functions (this is because the left hand side is linear in u). The second concept is that of separation of variables which consists of seeking product solutions of (2.2) of the form $X(x)Y(y)$. The basic technique in subsequent problems is essentially the determination of combinations of product solutions which are such that they satisfy all the conditions imposed. This approach is generally known as Fourier's method and is illustrated in the succeeding problems.

Problem 2.1 Solve the equation $\partial^2 u/\partial x^2 + \partial^2 u/\partial y^2 = 0,\ 0 < x < c,$ $0 < y < d,$ under the boundary conditions $u(0, y) = u(c, y) = 0, 0 < y < d,$ $u(x, 0) = f(x),\ u(x, d) = 0,\ 0 < x < c,$ where (i) $f(x) = \sin \pi x/c$, (ii) $f(x) = x(c - x)$.

(One physical interpretation of this problem would be the determination of the steady state temperature in a rectangular region where three sides of the rectangle are kept at zero temperature whilst the fourth is maintained at a known temperature $f(x)$.)

Solution. Assume u is of the form $X(x)Y(y)$ where X and Y are functions of x and y respectively. Substitution of this in the equation for u gives

$$\frac{X''}{X} = -\frac{Y''}{Y},$$

the dashes denoting derivatives with respect to the relevant variable. A function of x can only be equal to a function of y if both are constant. Hence $X'' = -\lambda X,\ Y'' = \lambda Y$, where λ is a constant (the separation constant). The condition $u(0, y) = u(c, y) = 0$ gives $X(0) = X(c) = 0$ and the general solution for X is $Ax + B, \lambda = 0; C \cos \lambda^{\frac{1}{2}}x + D \sin \lambda^{\frac{1}{2}}x, \lambda \neq 0$. For $\lambda = 0$ the conditions $X(0) = X(c) = 0$ show that $A = B = 0$ and, for $\lambda \neq 0, X(0) = 0$ gives $C = 0$ and $X(c) = 0$ then requires that $D \sin \lambda^{\frac{1}{2}}c = 0$.

23

Thus in order to have non-trivial solutions (i.e. $D \neq 0$)$c\lambda^{\frac{1}{2}}$ must be equal to $n\pi$, where n can be any integer. X will then be proportional to $\sin n\pi x/c$ and the equation for Y can be solved showing that Y will be an arbitrary linear combination of $\cosh n\pi y/c$, $\sinh n\pi y/c$. Thus possible solutions for u, satisfying the conditions at $x = 0$ and $x = c$, are $[A_n \cosh n\pi y/c + B_n \sinh n\pi y/c]\sin n\pi x/c$ where n is any integer and the arbitrary constants A_n, B_n may depend on n.

In accordance with the principle of superposition we thus consider the general form

$$u = \sum_{n=1}^{\infty} \left[A_n \cosh \frac{n\pi y}{c} + B_n \sinh \frac{n\pi y}{c} \right] \sin \frac{n\pi x}{c}$$

and attempt to find A_n and B_n to satisfy the conditions at $y = 0$ and $y = d$. The condition on $y = d$ will be satisfied if

$$A_n \cosh \frac{n\pi d}{c} = - B_n \sinh \frac{n\pi d}{c}$$

and, for case (i), the condition on $y = 0$ gives

$$\sin \frac{\pi x}{c} = \sum_{n=1}^{\infty} A_n \sin \frac{n\pi x}{c}, \qquad 0 < x < c.$$

Thus the A_n are the coefficients of the Fourier sine series in $[0, c]$ of $\sin \pi x/c$; this latter series is the function itself [cf. Problem 1.3] and hence $A_1 = 1$, $A_n = 0$, $n \neq 1$ and $u = - \operatorname{cosech}(\pi d/c) \sinh \pi(y - d)/c \sin \pi x/c$. It can be verified directly that this satisfies all the conditions imposed.

The only difference in case (ii) is that A_n will now be the coefficients of the sine series representing $x(c - x)$ in $[0, c]$, i.e.

$$A_n = \frac{2}{c} \int_0^c x(c - x) \sin \frac{n\pi x}{c} \, dx = \frac{4c^2}{n^3 \pi^3} [1 - \cos n\pi],$$

on integration by parts. Thus $A_n = 0$ (n even) and $A_{2r-1} = [8c^2/\pi^3(2r-1)^3]$, $r = 1, 2, \ldots$. The complete solution is thus

$$u = - \frac{8c^2}{\pi^3} \sum_{r=1}^{\infty} \frac{1}{(2r-1)^3} \cdot \operatorname{cosech} \frac{(2r-1)\pi d}{c} \sinh \frac{(2r-1)\pi(y-d)}{c}$$
$$\times \sin \frac{(2r-1)\pi x}{c}.$$

The fact that each term of an infinite series satisfies a differential equation does not necessarily mean that the sum satisfies this equation, as this would imply that differentiation of the series produces another convergent series. This is not always the case [cf. remarks following

Problem 1.16]. In the above series, however, the coefficients tend to zero sufficiently quickly for the series to be differentiated twice and the sum does satisfy Laplace's equation. The boundary conditions are all satisfied (by construction) and hence the above series represents the solution for case (ii). □

Problem 2.2 Solve the previous boundary value problem when $f(x) = x(c-x)$ and the condition on $x = 0$ is replaced by $u(0, y) = y(d-y)$, $0 < y < d$.

Solution. The boundary conditions on neither $x =$ constant nor $y =$ constant are homogeneous but, as Laplace's equation is linear, u can be written as $u_1 + u_2$, where u_1 and u_2 both satisfy Laplace's equation and satisfy homogeneous conditions in one or other of the variables. We have

$$u_1(0, y) = u_1(c, y) = 0, \qquad 0 < y < d,$$
$$u_1(x, 0) = x(c-x), u_1(x, d) = 0, \qquad 0 < x < c,$$
$$u_2(x, 0) = u_2(x, d) = 0, \qquad 0 < x < c,$$
$$u_2(0, y) = y(d-y), u_2(c, y) = 0, \qquad 0 < y < d.$$

u_1 is the solution in Problem 2.1, case (ii). The problem for u_2 is very similar to that for u_1 except that homogeneous boundary conditions are prescribed on $y = 0$ and $y = d$. A solution could be obtained by a very similar procedure to that of Problem 2.1 but in this particular case all additional work can be avoided by noticing that interchanging x and y and c and d reduces the problem for u_2 to that for u_1. Thus the solution can be obtained from the previous result as a sum of two infinite series.

This technique of splitting the problem into two separate ones with homogeneous conditions to be satisfied on one family of parallel lines in each case may be employed for any linear elliptic equation when non-homogeneous conditions are prescribed on the boundary of a rectangular region. □

Problem 2.3 Solve the equation

$$\frac{\partial^2 u}{\partial x^2} = \frac{1}{v^2} \frac{\partial^2 u}{\partial t^2}, \qquad 0 < x < l, t > 0,$$

where v is a constant, under the conditions

$$u(0, t) = 0, \quad \partial u(x, t)/\partial x = 0, \quad x = l; \quad t > 0,$$
$$u(0, x) = x, \quad \partial u(x, t)/\partial t = 0, \quad t = 0; \quad 0 < x < l.$$

(This problem can be interpreted physically as determining the longitudinal vibrations of a rod of length l clamped at $x = 0$ and free at $x = l$.

25

Initially the longitudinal displacement at a distance x from the fixed end is equal to x and the rod is at rest. These initial conditions could be produced by applying a force at the free end parallel to the rod and removing this force at time $t = 0$.)

Solution. Assuming a product solution $X(x)T(t)$ and proceeding as in Problem 2.1 gives $X'' = -\lambda X$, $T'' = -\lambda v^2 T$. The conditions on X are $X(0) = X'(l) = 0$ and the general form of X will be that of Problem 2.1. For $\lambda = 0$ the conditions $X(0) = X'(l) = 0$ again give $A = B = 0$; for $\lambda \neq 0$, $X(0) = 0$ gives $C = 0$ whilst $X'(l) = 0$ requires that $D \cos \lambda^{\frac{1}{2}} l = 0$. Thus for a non-trivial solution $\lambda^{\frac{1}{2}} l = (2n-1)\pi/2$, $n = 1, 2, \ldots$, where n is any integer and the corresponding X is proportional to $\sin (2n-1)\pi x/2l$. Hence solving the equation for T and using the principle of superposition suggests the general form

$$u = \sum_{n=1}^{\infty} \left[A_n \cos \frac{(2n-1)\pi v t}{2l} + B_n \sin \frac{(2n-1)\pi v t}{2l} \right] \sin \frac{(2n-1)\pi x}{2l}.$$

The condition $\partial u/\partial t = 0$, $t = 0$, is satisfied by taking $B_n = 0$ and $u(x, 0) = x$ implies that $x = \sum_{n=1}^{\infty} A_n \sin(2n-1)\pi x/2l$. Thus

$$A_n = \frac{2}{l} \int_0^l x \sin \frac{(2n-1)\pi x}{2l} \, dx = \frac{8l(-1)^{n+1}}{\pi^2(2n-1)^2}$$

and
$$u = \frac{8l}{\pi^2} \sum_{n=1}^{\infty} \frac{(-1)^{n+1}}{(2n-1)^2} \cos \frac{(2n-1)\pi v t}{2l} \sin \frac{(2n-1)\pi x}{2l}.$$

u satisfies the boundary conditions by construction, but as it is not permissible to differentiate twice under the summation sign though the separate terms satisfy the wave equation, it is not possible to show the series also satisfies the equation. u can be written as $F(x-vt)+F(x+vt)$ where

$$F(x) = \frac{4l}{\pi^2} \sum_{n=1}^{\infty} \frac{(-1)^{n+1}}{(2n-1)^2} \sin \frac{(2n-1)\pi x}{2l}.$$

$F(x)$ is related to the function sketched in Fig. 8 and is in fact the sine series of period $4l$ representing x in $[0, l]$. By analogy with Fig. 8 it has a discontinuous derivative at $x = \frac{1}{2}l, \frac{3}{2}l, \ldots$, and its second derivative does not exist at these points. Thus, as any twice differentiable function of $x \pm vt$ satisfies the wave equation, $F(x \pm vt)$ are solutions of the wave equation for a given t at all but a finite number of points.

The solution obtained is a solution in a somewhat generalized sense and in the general nomenclature of hyperbolic equations is called a 'weak' solution. The discontinuity in the derivative of F essentially reflects the discontinuity inherent in the initial conditions. $\square$

Simplified procedure In both Problems 2.1 and 2.3 the equation for X reduced to the simple harmonic one and it can be seen from these problems that this equation will always be obtained when the variable x only occurs explicitly in the equation as a constant multiple of $\partial^2 u/\partial x^2$. The same situation will also occur if there is a constant multiple of u in the equation. Also, by analogy with the previous problems, possible solutions X can be obtained for those cases when u or $\partial u/\partial x$ vanish at $x = 0, c$. The forms for two cases have already been found and the results for the other cases are

(i) $\partial u/\partial x = 0, x = 0, c$: $X = \cos n\pi x/c, n = 0, 1, 2, \ldots,$

(ii) $\partial u/\partial x = 0, x = 0; u = 0, x = c$:

$$X = \cos(2n-1)\pi x/2c, n = 1, 2, \ldots.$$

Thus for equations of the form of (2.2) with $a_2 = 0$ and a_1, a_3 constant it is possible to predict, for certain boundary conditions, without going through the general separation of variables procedure, the general form of the solution. It is useful to have the appropriate representations in the form of a table.

Table 1. *Representations for case* $a_2 = 0, a_1, a_3$ *constant*

Boundary conditions	Representation for u
(i) $u = 0, x = 0, x = c$	$\sum_{n=1}^{\infty} Y_n(y)\sin n\pi x/c$
(ii) $\partial u/\partial x = 0, x = 0, x = c$	$\sum_{n=0}^{\infty} Y_n(y)\cos n\pi x/c$
(iii) $u = 0, x = 0; \partial u/\partial x = 0, x = c$	$\sum_{n=1}^{\infty} Y_n(y)\sin(2n-1)\pi x/2c$
(iv) $\partial u/\partial x = 0, x = 0; u = 0, x = c$	$\sum_{n=1}^{\infty} Y_n(y)\cos(2n-1)\pi x/2c$

Entries (i) and (iii) respectively would have been appropriate for Problems 2.1 and 2.3. Once the appropriate representation has been chosen from the table then substituting this in the relevant equation and equating the coefficients of the trigonometric functions to zero (cf. remarks in Problem

27

1.3) gives the equation for Y_n. (Theorems 6 and 7 show that, provided u satisfies the relevant homogeneous conditions, the series can be differentiated term by term twice.) The approach is illustrated in the following problems.

Problem 2.4 Solve the equation $\partial^2 u/\partial x^2 = \partial u/\partial t$, $0 < x < l$, $t > 0$, under the boundary conditions $\partial u(x,t)/\partial x = 0$, $x = 0$; $u(l,t) = 0$, $t > 0$; $u(x,0) = \sin(\pi x/l)$, $0 < x < l$. (Physically this corresponds to calculating the temperature in the slab $0 < x < l$ when the end $x = 0$ is insulated whilst the end $x = l$ is maintained at zero temperature, the initial temperature being $\sin(\pi x/l)$.)

Solution. Table 1(iv) gives the appropriate representation as $u = \sum_{n=1}^{\infty} T_n(t)\cos(2n-1)\pi x/2l$. Substituting this into the equation gives

$$-\sum_{n=1}^{\infty} \frac{(2n-1)^2 \pi^2}{4l^2} T_n(t)\cos\frac{(2n-1)\pi x}{2l} = \sum_{n=1}^{\infty} \frac{dT_n}{dt}\cos\frac{(2n-1)\pi x}{2l},$$

equating the coefficients of $\cos(2n-1)\pi x/2l$ in this equation gives

$$\frac{dT_n}{dt} = -\frac{(2n-1)^2 \pi^2}{4l^2} T_n,$$

and the general solution is $T_n = A_n \exp[-(2n-1)^2\pi^2 t/4l^2]$.

The boundary condition at $t = 0$ shows that $T_n(0)$ ($= A_n$) are the Fourier coefficients in the expansion of $\sin \pi x/l$ as a cosine series of period $4l$. This was considered in Problem 1.15 and A_n can be found from that problem by replacing c by l. Hence

$$u = -\frac{8}{\pi}\sum_{n=1}^{\infty}\frac{\exp[-(2n-1)^2\pi^2 t/4l^2]}{(2n+1)(2n-3)}\cos\frac{(2n-1)\pi x}{2l}.$$

The coefficients in this series are exponentially damped for large n. Thus the order of differentiation and summation can be interchanged and u satisfies the differential equation; u also satisfies the boundary conditions (by construction). □

Problem 2.5 Obtain a solution u of Laplace's equation finite everywhere in the sector $0 \leqslant r \leqslant a$, $0 \leqslant \theta \leqslant \alpha$, where r, θ are polar coordinates, and such that it has zero normal derivative on the lines $\theta = 0$ and $\theta = \alpha$ and is equal to θ on $r = a$.

Solution. u satisfies $r^2 \partial^2 u/\partial r^2 + r\,\partial u/\partial r + \partial^2 u/\partial\theta^2 = 0$, which is of the form of equation (2.2) with x replaced by θ. Table 1(ii) gives the appropriate representation for u to be $\sum_{n=0}^{\infty} F_n(r)\cos n\pi\theta/\alpha$. Substituting this in the equation and equating the coefficients of $\cos n\pi\theta/\alpha$ to zero gives

$$r^2 F_n'' + rF_n' - (n^2\pi^2/\alpha^2)F_n = 0.$$

For $n \neq 0$, independent solutions of this equation are $r^{\pm n\pi/\alpha}$ and the finiteness condition shows that the positive sign must be chosen. For $n = 0$ the general solution is $A \ln r + B$ and for finiteness $A = 0$. The appropriate general solution for F_n is thus $A_n r^{n\pi/\alpha}$ (all n) and the boundary condition on $r = a$ shows that $A_n a^{n\pi/\alpha}$ are the Fourier coefficients of the cosine series of θ in $[0, \alpha]$, i.e.

$$A_0 = \frac{1}{\alpha} \int_0^\alpha \theta \, d\theta = \frac{1}{2}\alpha, \qquad A_n' a^{n\pi/\alpha} = \frac{2}{\alpha} \int_0^\alpha \theta \cos \frac{n\pi\theta}{\alpha} \, d\theta.$$

Integration shows that $A_n = 0$ (n even) and $A_{2r-1} = -4\alpha/\pi^2(2r-1)^2$, $[r = 1, 2, \ldots]$ (cf. Problem 1.10(a)). The infinite series for the solution can now be written down. $\qquad\square$

The above analysis may be generalized to the case where u satisfies the equation $(\nabla^2 + k^2)u = 0$ where k is a real constant, and the boundary conditions are unchanged. The equation for F_n is

$$\alpha^2 r^2 F_n'' + \alpha^2 rF_n' + (k^2\alpha^2 r^2 - n^2\pi^2)F_n = 0,$$

or

$$\frac{d^2 F_n}{dz^2} + \frac{1}{z}\frac{dF_n}{dz} + \left(1 - \frac{n^2\pi^2}{\alpha^2 z^2}\right)F_n = 0,$$

where $z = kr$. This is Bessel's equation of order $n\pi/\alpha$, and the solution finite for $z = 0$ is $J_{n\pi/\alpha}(z)$ which is the Bessel function of order $n\pi/\alpha$. Thus the solution of this second problem is found by replacing $r^{n\pi/\alpha}$, $a^{n\pi/\alpha}$ in the solution found earlier by $J_{n\pi/\alpha}(kr)$, $J_{n\pi/\alpha}(ka)$.

Problem 2.6 Find a function satisfying Laplace's equation, finite and continuous everywhere in the circle $0 \leqslant r \leqslant a$, and equal to $\cos^3\theta$ on the circumference, r, θ being polar coordinates.

Solution. There are no boundary conditions on $\theta = $ constant which would let us employ Table 1 but assuming a solution $F(r)G(\theta)$ gives

$$G'' = -\lambda G, \qquad \frac{1}{r}\frac{d}{dr}\left(r\frac{dF}{dr}\right) = \lambda F.$$

G has the general solution $A \cos \lambda^{\frac{1}{2}}\theta + B \sin \lambda^{\frac{1}{2}}\theta$ and λ can be found by noticing that G will not be continuous and single valued in $[0, 2\pi]$ unless $\lambda = n^2$ (n any integer). The solution F_n corresponding to $\lambda = n^2$ can be obtained from Problem 2.5 on setting $\alpha = \pi$; the finiteness condition shows that F_n will be proportional to r^n. Thus the general form for the solution will be

$$A_0 + \sum_{n=1}^{\infty} r^n (A_n \cos n\theta + B_n \sin n\theta).$$

Setting $r = a$ shows that A_n, B_n can be calculated in terms of the Fourier coefficients of $\cos^3\theta$ and the latter are most easily calculated from $\cos 3\theta = 4\cos^3\theta - 3\cos\theta$. Hence $B_n \equiv A_n \equiv 0$, $n \neq 1, 3$, $A_1 a = \frac{3}{4}$, $A_3 a^3 = \frac{1}{4}$. $\square$

Problem 2.7 Find u satisfying $\nabla^2 u = \partial u/\partial t$, $t > 0$, finite everywhere within the sphere $r \leqslant a$, vanishing on $r = a$ and equal to unity when $t = 0$.

(This problem is equivalent to the determination of the temperature in a spherical region, the boundary of which is maintained at zero temperature, the initial temperature being prescribed.)

Solution. The boundary conditions are spherically symmetric and it is reasonable to assume that u is a function of r only, hence

$$\frac{1}{r^2} \frac{\partial}{\partial r} \left(r^2 \frac{\partial u}{\partial r} \right) = \frac{\partial u}{\partial t}.$$

Writing ru as v gives $\partial^2 v/\partial r^2 = \partial v/\partial t$, and in order that u be finite when $r = 0$ it is necessary for v to vanish at $r = 0$. v will also vanish on $r = a$ and, as v satisfies an equation where r only occurs in the term $\partial^2 v/\partial r$. Table 1(i) shows that the appropriate representation for v is $\sum_{n=1}^{\infty} T_n(t)\sin n\pi r/a$, and substituting this in the equation gives $T_n' = (n^2\pi^2/a^2)T_n$, with a general solution $T_n = A_n \exp(-n^2\pi^2 t/a^2)$. The condition at $t = 0$ shows that A_n are the Fourier coefficients of the sine series of r on $[0, a]$, i.e. $A_n = (2/a)\int_0^a r \sin(n\pi r/a)\, dr = 2(-1)^{n+1}a/n\pi$. $\square$

Problem 2.8 Solve the equation $\partial^2 u/\partial x^2 + \partial^2 u/\partial y^2 = \partial^2 u/\partial t^2$, $t > 0$, $0 < x < c$, $0 < y < d$, under the conditions $u = 0$, $x = 0$, $x = c$; $y = 0$, $y = d$; $u = xy(c-x)(d-y)$, $\partial u/\partial t = 0$, $t = 0$, $0 < x < c$, $0 < y < d$.

(A physical interpretation of this is the determination of the shape of a rectangular membrane given its initial shape and that its initial velocity is zero.)

Solution. As this problem involves three independent variables Table 1 is not directly applicable, but assuming a product solution gives

$$\frac{X''}{X} + \frac{Y''}{Y} = \frac{T''}{T}.$$

Both sides of this equation must be constant i.e.

$$\frac{X''}{X} + \frac{Y''}{Y} = -\lambda, \qquad \frac{T''}{T} = -\lambda.$$

As X and Y are functions of independent variables x and y, both the terms in the first of the latter equations must be separately constant, i.e.

$$\frac{Y''}{Y} = -\mu, \qquad \frac{X''}{X} = \mu - \lambda.$$

The equation for Y is similar to the equation of Problem 2.1 and in order that u vanishes at $y = 0, d$ it is necessary that $\mu = m^2\pi^2/d^2$ (m an integer). Similarly from the equation for X and the conditions at $x = 0, c$ it is found that $(\lambda - \mu) = n^2\pi^2/c^2$ (n an integer). Substituting λ, μ in the equation for T shows that the general form for u is

$$\sum_{n=1}^{\infty} \sum_{m=1}^{\infty} [A_{mn} \cos K_{mn} t + B_{mn} \sin K_{mn} t] \sin\frac{n\pi x}{c} \sin\frac{m\pi y}{d},$$

where $K_{mn}^2 = n^2\pi^2/c^2 + m^2\pi^2/d^2$. $\partial u/\partial t$ will vanish for $t = 0$ if $B_{mn} = 0$ and the other condition at $t = 0$ gives

$$xy(c-x)(d-y) = \sum_{n=1}^{\infty} \sum_{m=1}^{\infty} A_{mn} \sin\frac{n\pi x}{c} \sin\frac{m\pi y}{d}, \quad 0 \leqslant x \leqslant c, 0 \leqslant y \leqslant d.$$

This is a double Fourier series and, for a general left-hand side, A_{mn} could be found by successive application of equation (1.5), i.e.

$$A_{mn} = \frac{4}{cd} \int_0^c \int_0^d xy(c-x)(d-y) \sin\frac{n\pi x}{c} \sin\frac{m\pi y}{d} \, dxdy.$$

This elaborate approach can be avoided in this case as the integrand is a product of a function of x and a function of y. The sine series for $x(c-x)$ has been obtained in Problem 2.1 and that for $y(d-y)$ can be derived from this by replacing x, c by y, d. Hence $A_{mn} = 0$ if either m and/or n is even and

$$A_{2r-1, 2s-1} = \frac{64c^2d^2}{\pi^6(2r-1)^3(2s-1)^3}, \qquad r, s = 1, 2, \ldots.$$

Hence

$$u = \frac{64c^2d^2}{\pi^6} \sum_{r=1}^{\infty} \sum_{s=1}^{\infty} \frac{\cos K_{2r-1, 2s-1} t}{(2r-1)^3(2s-1)^3} \sin\frac{(2r-1)\pi y}{d} \sin\frac{(2s-1)\pi x}{c}. \qquad \square$$

Problem 2.9   Solve the equation

$$\frac{\partial^4 u}{\partial x^4} + \frac{1}{a^4}\frac{\partial^2 u}{\partial t^2} = 0, \qquad 0 < x < l, t > 0$$

where a is constant, under the conditions $u = \partial^2 u/\partial x^2 = 0$, $x = 0$, $x = l$, $t > 0$; $u(x,0) = f(x)$, $\partial u(x,t)\,\partial t = 0$, $t = 0$, $0 < x < l$.

(This can be interpreted physically as determining the transverse displacement of an elastic beam of length l, simply supported at its ends. The initial velocity of the beam is zero and its displacement is $f(x)$.)

Solution. The equation is not of the form of (2.2) but assuming a product solution $X(x)T(t)$ gives $X'''' - \lambda X = 0$, $T'' + \lambda a^4 T = 0$, where λ is the separation constant. Making the trial substitution $X = Ae^{\mu x}$ gives $\mu^4 = \lambda$, i.e. $\mu = \pm\lambda^{\frac{1}{4}}$, $\pm i\lambda^{\frac{1}{4}}$ hence the general solution is

$$X = A\cos\lambda^{\frac{1}{4}}x + B\sin\lambda^{\frac{1}{4}}x + C\cosh\lambda^{\frac{1}{4}}x + D\sinh\lambda^{\frac{1}{4}}x.$$

The conditions at $x = 0$ give $A + C = 0 = A - C$; i.e. $A = C = 0$; the conditions at $x = l$ give

$$B\sin\lambda^{\frac{1}{4}}l + D\sinh\lambda^{\frac{1}{4}}l = 0 = D\sinh\lambda^{\frac{1}{4}}l - B\sin\lambda^{\frac{1}{4}}l.$$

Thus, if B and D are not to vanish, either $\sinh\lambda^{\frac{1}{4}}l = 0$ or $\sin\lambda^{\frac{1}{4}}l = 0$. The former gives $\lambda = 0$ (purely imaginary values for $\lambda^{\frac{1}{4}}$ are roots of $\sin\lambda^{\frac{1}{4}}l = 0$) which implies $X = 0$; hence $\lambda^{\frac{1}{4}} = n\pi/l$ where n is any integer. Hence $D = 0$, and the appropriate representation for u is thus

$$\sum_{n=1}^{\infty} T_n(t)\sin n\pi x/l \text{ where } T_n'' + (n^4\pi^4 a^4/l^4)\,T_n = 0.$$ The general solution for T_n is $A_n\cos n^2\pi^2 a^2 t/l^2 + B_n\sin n^2\pi^2 a^2 t/l^2$, $\partial u/\partial t = 0$ at $t = 0$ implies $B_n = 0$ and $u(0,t) = f(x)$ implies that A_n are the Fourier coefficients of the sine series for $f(x)$ in $[0, l]$, i.e.

$$A_n = \frac{2}{l}\int_0^l f(x)\sin\frac{n\pi x}{l}\,dx.$$

Thus, for any given $f(x)$, a complete series solution has been obtained. $\square$

2.3 Inhomogeneous Problems It is necessary, in order to solve equation (2.1) for $f \not\equiv 0$ with u or $\partial u/\partial x$ taking prescribed but non-zero values on $x = 0, c$, to modify the techniques used in the previous problems.

In some cases f and/or the form of the boundary conditions on $x = 0, c$ are sufficiently simple for a function U (often a polynomial in x) to be found by inspection, which satisfies both the equations and the conditions on $x = 0, c$. It is unlikely that a simple function U can be found satisfying whatever conditions are prescribed on $y = $ constant also, but writing the unknown u as $U + v$ will give an equation of the form (2.2) for v with v satisfying homogeneous conditions on $x = 0, c$. The problem for v will be of the general type discussed earlier.

Homogeneous boundary conditions For problems where u or $\partial u/\partial x$ are prescribed to vanish for $x = 0, c$ (but it is not possible to use the above simple approach), it is only necessary to modify the previous approach very slightly. If the problem for $f \equiv 0$ can be solved by using a representation for u as a Fourier series, then a useful method to try for the inhomogeneous problem is to assume a Fourier series in x of the same type for f. Substituting both forms into the relevant partial differential equation and equating coefficients of trigonometric functions enables a differential equation to be obtained for the Fourier coefficients. Both the above methods are illustrated in the following problems.

Problem 2.10 Find a function u satisfying the equation and conditions of Problem 2.3 except that at $x = l$ which is replaced by $\partial u/\partial x = 1$.

Solution. x satisfies both the differential equation and the conditions at $x = 0, l$. Thus writing u as $x + v$ shows that v satisfies the same differential equation, and the boundary conditions on v at $x = 0$ and $x = l$ are those imposed in Problem 2.3. The conditions $u = \partial u/\partial t = 0$ at $t = 0$ give $v = -x, \partial v/\partial t = 0$ at $t = 0$. Hence $-v$ is the function obtained in Problem 2.3 and

$$u = x - \frac{8l}{\pi^2} \sum_{n=1}^{\infty} \frac{(-1)^{n+1}}{(2n-1)^2} \cos(2n-1)\frac{\pi vt}{2l} \sin(2n-1)\frac{\pi x}{2l}. \qquad \square$$

Problem 2.11 Solve $\partial^2 u/\partial x^2 + \partial^2 u/\partial y^2 = 2$, $0 < x < c$, $0 < y < d$, under the boundary conditions $u = 0$ on $x = 0, c$; $0 < y < d$; $u = 0$, $y = 0$, $\partial u/\partial y = 0$, $y = d$, $0 < x < c$.

Solution. The right hand side of the equation is sufficiently simple to suggest seeking elementary solutions. Particular solutions are x^2 and y^2, and adding to either linear combinations of x and y will still produce solutions of the differential equation. If we consider $x^2 + ax + b$ then the conditions at $x = 0$ and $x = c$ are satisfied by $b = 0$, $a = -c$. Hence writing u as $x^2 - cx + v$ shows that v is a solution of Laplace's equation vanishing on $x = 0, c$; the conditions on u at $y = 0, d$ show that $v = cx - x^2$, $y = 0$; $\partial v/\partial y = 0$, $y = d$. The appropriate representation for v is that of Problem 2.1, i.e.

$$v = \sum_{n=1}^{\infty} [A_n \cosh n\pi y/c + B_n \sinh n\pi y/c]\sin n\pi x/c$$

The conditions on $y = d$ will be satisfied if $A_n \sinh n\pi d/c = -B_n \cosh n\pi d/c$ and that on $y = 0$ shows that A_n are the coefficients of the sine series of $x(c-x)$ in $[0, c]$ and thus A_n have the same values as in Problem 2.1(ii).

33

Hence

$$u = x^2 - cx + \frac{8c^2}{\pi^3} \sum_{r=0}^{\infty} \frac{1}{(2r+1)^3} \operatorname{sech} \frac{(2r+1)\pi d}{c} \cosh \frac{(2r+1)\pi(y-d)}{c}$$
$$\sin \frac{(2r+1)\pi x}{c}$$

An alternative approach would be to satisfy the boundary conditions on $y = 0$ and $y = d$; $y^2 + ay + b$ satisfies both the equation and these conditions if $a = -2d$ and $b = 0$. Hence writing u as $y^2 - 2dy + w$ shows that w satisfies Laplace's equation and the appropriate homogeneous conditions on $y = 0, d$. On $x = 0$ and $x = c, w$ is equal to $2dy - y^2$. The boundary value problem for w can be solved by means of a representation of the form of Table 1(iii). $\qquad\square$

Problem 2.12 Solve $\partial^2 u/\partial x^2 - \partial^2 u/\partial t^2 = \exp(-t)$, $t > 0$, $0 < x < l$. under the conditions $u(0, t) = u(l, t) = 0$, $t > 0$; $u(x, t) = \partial u(x, t)/\partial t = 0$, $t = 0, 0 < x < l$.

Solution. It is not easy to find by inspection a function satisfying both the equation and the boundary conditions and we need to use the more general approach described. If the right-hand side of the equation were zero then the appropriate representation would be $u = \sum_{n=1}^{\infty} T_n(t) \sin n\pi x/l$ and we therefore need to obtain a sine series representation for $\exp(-t)$ in $[0, l]$. This representation will of course be $\exp(-t)$ times the sine series for unity, the latter series can be calculated by the methods of §1.2, and it is found that

$$1 = \frac{4}{\pi} \sum_{r=1}^{\infty} \frac{\sin(2r-1)\pi x/l}{2r-1}, \qquad 0 < x < l.$$

Thus substituting this series and that for u in the above differential equation gives

$$\sum_{n=1}^{\infty} \left[-\frac{n^2\pi^2}{l^2} T_n - T_n'' \right] \sin \frac{n\pi x}{l} = \frac{4}{\pi} \sum_{r=1}^{\infty} \frac{\sin(2r-1)\pi x/l}{2r-1} \exp(-t)$$

Equating coefficients of the trigonometric functions gives

$$-\frac{4r^2\pi^2}{l^2} T_{2r} - T_{2r}'' = 0, \qquad r = 0, 1, 2, \ldots,$$

$$-\frac{(2r-1)^2\pi^2}{l^2} T_{2r-1} - T_{2r-1}'' = \frac{4\exp(-t)}{\pi(2r-1)}, \qquad r = 1, 2, \ldots.$$

$u = \partial u/\partial t = 0$ at $t = 0$ imply that $T_n = T_n' = 0$, for all n when $t = 0$, and as the equation and the conditions for T_{2r} are both homogeneous it follows that $T_{2r} \equiv 0$. The general solution for T_{2r-1} is

$$T_{2r-1} = -\frac{4l^2 \exp(-t)}{\pi(2r-1)[(2r-1)^2\pi^2 + l^2]} + A_{2r-1}\cos\frac{(2r-1)\pi t}{l}$$
$$+ B_{2r-1}\sin\frac{(2r-1)\pi t}{l}.$$

The conditions at $t = 0$ give

$$A_{2r-1} = \frac{4l^2}{\pi(2r-1)[l^2 + (2r-1)^2\pi^2]},$$

$$B_{2r-1} = -\frac{4l^3}{\pi^2(2r-1)^2[l^2 + (2r-1)^2\pi^2]}.$$

Substitution of these values into the expression for T_{2r-1} gives

$$u = -\frac{4l^2}{\pi}\sum_{r=1}^{\infty}\frac{1}{(2r-1)[l^2 + (2r-1)^2\pi^2]}\left[\exp(-t) - \cos\frac{(2r-1)\pi t}{l}\right.$$
$$\left.+ \frac{l}{\pi(2r-1)}\sin\frac{(2r-1)\pi t}{l}\right]\sin\frac{(2r-1)\pi x}{l}. \qquad \square$$

Problem 2.13 Solve $\partial^2 u/\partial x^2 - \partial u/\partial t = \exp(-t)\sin x$, $0 < x < c$, $t > 0$, under the conditions $u(0, t) = 0$, $\partial u(x, t)/\partial x = 0$, $x = c$; $t > 0$, $u(x, 0) = 0$, $0 < x < c$.

Solution. If the right hand side of the above equation were zero it follows from Table 1(iii) that an appropriate representation for u would be $\sum_{n=1}^{\infty}T_n(t)\sin(2n-1)\pi x/2c$, and we therefore require a similar representation for $\sin x$. This is obtained as in § 1.3, and

$$\sin x = 8c\cos c\sum_{n=1}^{\infty}\frac{(-1)^{n+1}}{[(2n-1)^2\pi^2 - 4c^2]}\sin\frac{(2n-1)\pi x}{2c}.$$

[It has been assumed that c is not equal to $(2n-1)\pi/2$ for any integer n; in this case the representation of $\sin x$ would be itself.]

Substituting both forms in the equation gives

$$\sum_{n=1}^{\infty}\left[-\frac{(2n-1)^2\pi^2}{4c^2}T_n - \frac{dT_n}{dt}\right]\sin\frac{(2n-1)\pi x}{2c} = 8c\cos c\exp(-t)$$
$$\sum_{n=1}^{\infty}\frac{(-1)^{n+1}}{(2n-1)^2\pi^2 - 4c^2}\sin\frac{(2n-1)\pi x}{2c}$$

and hence

$$\frac{dT_n}{dt}+\frac{(2n-1)^2\pi^2}{4c^2}T_n=\frac{8c\cos c(-1)^n\exp(-t)}{(2n-1)^2\pi^2-4c^2}.$$

The general solution of this equation is

$$T_n=A_n\exp[-(2n-1)^2\pi^2t/4c^2]+\frac{32c^3\cos c(-1)^n\exp(-t)}{[(2n-1)^2\pi^2-4c^2]^2}$$

and to satisfy the condition on $t=0$ we require that

$$A_n=-\frac{32c^3\cos c(-1)^n}{[(2n-1)^2\pi^2-4c^2]^2}.$$

Hence a complete series solution for u may be written down. □

Problem 2.14 Solve $\partial^2u/\partial x^2=\partial u/\partial t$, $0<x<l$, $t>0$, under the conditions $\partial u(x,t)/\partial x=q$, where q is constant, for $x=0$, $u(l,t)=T_0$ where T_0 is another constant, and $u(x,0)=0$.

Solution. Any linear function of x satisfies the equation and the particular choice $q(x-l)+T_0$ also satisfies the boundary conditions. Writing u as $q(x-l)+T_0+v$ shows that v satisfies the same equation as u, and also satisfies homogeneous condition for $x=0$, and $x=+l$ and $v(x,0)=-T_0-q(x-l)$. The appropriate representation to choose for v will be that of Problem 2.4 and the results can be obtained from those of that problem by replacing $\sin\pi x/l$ by $-T_0-q(x-l)$; this means that

$$v=\sum_{n=1}^{\infty}A_n\exp[-(2n-1)^2\pi^2t/4l^2]\cos\frac{(2n-1)\pi x}{2l},$$

where

$$A_n=-\frac{2}{l}\int_0^l[T_0+q(x-l)]\cos(2n-1)\pi x/2l\,dx=\frac{4(-1)^nT_0}{(2n-1)\pi}+\frac{8lq}{(2n-1)^2\pi^2}.$$ □

Inhomogeneous boundary conditions When the prescribed values of u or $\partial u/\partial x$ on $x=0,c$ are not zero and the problem cannot be reduced to homogeneous form, then a general, though standard, approach has to be used. If the corresponding homogeneous problem could be solved by using a particular Fourier series representation then the technique is to multiply the differential equation by the general member of the Fourier series [e.g. $\sin n\pi x/c$ for the case of Table 1(i)] and integrate the equation with respect to x from $x=0$ to $x=c$. Integration by parts of the term involving $\partial^2u/\partial x^2$, [for the case when Fourier series are applicable there will be no term involving $\partial u/\partial x$], will lead to a differential equation for the

36

Fourier coefficients of u. This technique is illustrated in the following problems.

Problem 2.15 Solve the equation of the previous problem under the conditions

$$u(0,t) = t, u(c,t) = 0, t > 0, u(x,0) = 0, 0 < x < c.$$

Solution. The corresponding homogeneous problem [i.e. $u(0,t) = u(c,t) = 0$] could be solved by a sine series representation and we thus multiply both sides of the differential equation by $\sin n\pi x/c$ and integrate with respect to x from 0 to c, i.e.

$$\int_0^c \frac{\partial^2 u}{\partial x^2} \sin \frac{n\pi x}{c} dx = \int_0^c \frac{\partial u}{\partial t} \sin \frac{n\pi x}{c} dx.$$

Integration by parts twice of the left-hand side of this equation gives

$$\left[\frac{\partial u}{\partial x} \sin \frac{n\pi x}{c} \right]_{x=0}^{x=c} - \frac{n\pi}{c} \left[u \cos \frac{n\pi x}{c} \right]_{x=0}^{x=c} - \frac{n^2\pi^2}{c^2} \int_0^c u \sin \frac{n\pi x}{c} dx$$

$$= \int_0^c \frac{\partial u}{\partial t} \sin \frac{n\pi x}{c} dx = \frac{\partial}{\partial t} \int_0^c u \sin \frac{n\pi x}{c} dx.$$

$u(x,t)$ can be represented in $0 < x < c$ as $\sum_{n=1}^{\infty} b_n(t) \sin n\pi x/c$, where

$$b_n = \frac{2}{c} \int_0^c u(x,t) \sin \frac{n\pi x}{c} dx.$$

Thus, on applying the conditions at $x = 0, c$ to the equation obtained by integration of parts,

$$\frac{db_n}{dt} + \frac{n^2\pi^2}{c^2} b_n = \frac{2n\pi t}{c^2},$$

i.e. the required equation for b_n. The general solution is

$$b_n = A_n \exp(-n^2\pi^2 t/c^2) + 2\left(\frac{t}{n\pi} - \frac{c^2}{n^3\pi^3} \right)$$

and in order to satisfy $u(x,0) = 0$, $A_n = 2c^2/n^3\pi^3$ giving

$$u = \sum_{n=1}^{\infty} \frac{2c^2}{n^3\pi^3} [\exp(-n^2\pi^2 t/c^2) - 1] \sin \frac{n\pi x}{c} + \frac{2t}{\pi} \sum_{n=1}^{\infty} \frac{1}{n} \sin \frac{n\pi x}{c}.$$

The last series is very slowly convergent and for computational purposes is not satisfactory, but its exact form is given in equation (1) Appendix 1 and use of this result and (7) of the same appendix considerably eases numerical computation. In general the results of Appendix 1 are often useful in avoiding numerical calculation of slowly convergent series.

It would not have been correct in this case to have differentiated the sine series twice to calculate the Fourier series of $\partial^2 u/\partial x^2$ as the sine series only represents u in the open interval and cannot be differentiated term by term [c.f. remarks preceding Theorem 6]. An alternative method would have been to use Theorem 6(ii) to calculate the cosine series of $\partial u/\partial x$ and then applied Theorem 6(ii) to this series to obtain the Fourier series for $\partial^2 u/\partial x^2$. The above technique involving integration by parts is more useful in that it avoids the necessity for remembering formulae.

It would also have been possible, with some ingenuity, to have reduced the problem for u to a homogeneous one. A simple function satisfying the boundary condition is $u_0 = t(1-x/c)$, but

$$\frac{\partial^2 u_0}{\partial x^2} - \frac{\partial u_0}{\partial t} = 1 - \frac{x}{c}$$

and it is therefore necessary to add to u_0 a function $f(x)$ such that $f'' = -1 + x/c$ and with $f(0) = 0 = f(c)$; such a function is $(x^3 - 3cx^2 + 2c^2 x)/6c$. Hence writing u as

$$t\left(1 - \frac{x}{c}\right) + \frac{1}{6c}(x^3 - 3cx^2 + 2c^2 x) + v$$

yields a homogeneous problem for v. The form in which u is obtained in this case is precisely that obtained by using the results of Appendix 1 to transform the series solution obtained previously. $\qquad\square$

Problem 2.16 Solve $\partial^2 u/\partial x^2 = \partial^2 u/\partial t^2$, $t > 0$, $0 < x < c$, subject to the conditions $u(0, t) = 0$, $\partial u(x, t)/\partial x = \sin \omega t, x = c, t > 0$; $u(x, 0) = \partial u(x, t)/\partial t = 0$, $t = 0$, $0 < x < c$ [ω is not an odd integer multiple of $\pi/2c$].

Solution. The appropriate representation for the homogeneous problem would be that of Table 1(iii) and we thus multiply both sides of the equation by $\sin (2n-1)\pi x/2c$ and integrate with respect to x from $x = 0$ to $x = c$, i.e.

$$\int_0^c \frac{\partial^2 u}{\partial x^2} \sin \frac{(2n-1)\pi x}{2c} \, dx = \int_0^c \frac{\partial^2 u}{\partial t^2} \sin \frac{(2n-1)\pi x}{2c} \, dx.$$

Integration by parts twice gives

$$\left[\frac{\partial u}{\partial x} \sin \frac{(2n-1)\pi x}{2c}\right]_{x=0}^{x=c} - \frac{(2n-1)\pi}{2c}\left[u \cos \frac{(2n-1)\pi x}{2c}\right]_{x=0}^{x=c}$$

$$- \frac{(2n-1)^2 \pi^2}{4c^2}\int_0^c u \sin \frac{(2n-1)\pi x}{2c} \, dx = \frac{\partial^2}{\partial t^2}\int_0^c u \sin \frac{(2n-1)\pi x}{2c} \, dx.$$

If b_n is defined by $u = \sum_{n=1}^{\infty} b_n(t)\sin(2n-1)\pi x/2c$ it follows, on using the prescribed conditions, that

$$\frac{d^2 b_n}{dt^2} + \frac{(2n-1)^2\pi^2 b_n}{4c^2} = \frac{2(-1)^{n+1}\sin \omega t}{c}.$$

The solution of this satisfying the conditions that $db_n/dt = b_n = 0$ at $t = 0$ is

$$\frac{8c(-1)^{n+1}}{(2n-1)^2\pi^2 - 4c^2\omega^2}\left[\sin \omega t - \frac{2\omega c}{(2n-1)\pi}\sin\frac{(2n-1)\pi t}{2c}\right].$$

Thus the series solution for u is obtained. $\qquad\square$

Problem 2.17 Solve the equation

$$\frac{\partial^4 u}{\partial x^4} + 2\frac{\partial^4 u}{\partial x^2 \partial y^2} + \frac{\partial^4 u}{\partial y^4} = 0, \quad 0 < x < c, 0 < y < d,$$

under the conditions $u(0, y) = u(c, y) = 0, (\partial^2 u/\partial x^2)_{x=0} = (\partial^2 u/\partial x^2)_{x=c} = 2, 0 < y < d; u(x, 0) = u(x, d) = 0, \partial u/\partial y = 0, y = 0, d, 0 < x < c.$

(This problem can be interpreted physically as calculating the deflection of an elastic plate, two of whose edges are clamped and the other two are simply supported. It is also acted upon by bending moments uniformly distributed along the free edges.)

Solution. This boundary value problem is of a type not previously considered and it can be verified that the method of separation of variables is not suitable. The problem can be somewhat simplified by making all the boundary conditions on $x = 0$ and $x = c$ homogeneous and this can be achieved by setting $u = x(x-c)+v$ and v will satisfy the same equation as u. The conditions on v show that it can be represented in $[0, c]$ by a series of the form $\sum_{n=1}^{\infty} Y_n(y)\sin n\pi x/c$ and, as $\partial^2 v/\partial x^2$ vanishes at $x = 0, c$, it can be shown from Theorem 6 that it can be represented as $-\sum_{n=1}^{\infty} (n^2\pi^2/c^2)Y_n \sin n\pi x/c$ for x in $[0, c]$. Further application of Theorem 6 shows that the series representing $\partial^4 v/\partial x^4$ is $\sum_{n=1}^{\infty} (n^4\pi^4/c^4)Y_n \sin n\pi x/c$. Substituting the series in the equation and equating coefficients of $\sin n\pi x/c$ gives

$$n^4\pi^4 Y_n - 2n^2\pi^2 c^2 Y_n'' + c^4 Y_n'''' = 0,$$

the conditions on Y_n are that $Y_n' = 0$ at $y = 0, d$ and $Y_n(0)$ and $Y_n(d)$ are

39

the coefficients of the Fourier sine series of $x(c-x)$. These have been obtained in Problem 2.1 and thus $Y_n(0) = Y_n(d) = 0$ for n even and

$$Y_{2r-1}(0) = Y_{2r-1}(d) = \frac{8c^2}{(2r-1)^3\pi^3}, \qquad r = 1,2,\dots.$$

The general solution for Y_n is a linear combination of $\cosh n\pi y/c$, $\sinh n\pi y/c$, $y\cos n\pi y/c$, $y\sin n\pi y/c$ and there will be four conditions to determine the four constants. The calculations can be simplified by noticing that Y_n satisfies the same condition at both boundaries; i.e. it is symmetric about $y = \frac{1}{2}d$ and we choose the linear combination with this property, i.e.

$$Y_n = A_n \cosh \frac{n\pi}{c}(y-\tfrac{1}{2}d) + B_n(y-\tfrac{1}{2}d)\sinh\frac{n\pi}{c}(y-\tfrac{1}{2}d).$$

For n even it follows from the conditions $Y_n(0) = Y_n(d) = Y_n'(0) = Y_n'(d)=0$ that $A_n = B_n = 0$, thus

$$A_{2r-1}\cosh\frac{(2r-1)\pi d}{2c} + B_{2r-1}\frac{d}{2}\sinh\frac{(2r-1)\pi d}{2c} = \frac{8c^2}{(2r-1)^3\pi^3},$$

$$\frac{(2r-1)\pi}{c}A_{2r-1}\sinh\frac{(2r-1)\pi d}{2c} + B_{2r-1}\sinh\frac{(2r-1)\pi d}{2c}$$

$$+ \frac{(2r-1)\pi d}{c}\cosh\frac{(2r-1)\pi d}{c} = 0.$$

These equations can now be solved to give A_{2r-1}, B_{2r-1} and the series solution for v is then found. $\qquad\qquad\square$

EXERCISES

1. Solve $\partial^2 u/\partial x^2 + \partial^2 u/\partial y^2 = 0$, $0 < x < a$, $0 < y < b$, under the conditions $u(0,y) = u(a,y) = 0$, $0 < y < b$; $u(x,0) = 0$, $\partial u/\partial y = q$, $y = b$, $0 < x < a$, where q is a constant.

2. Obtain a solution $u(\rho,\theta)$ of Laplace's equation finite everywhere within the circle of radius a, (ρ,θ) being the usual polar coordinates, for the two cases

   (i) $u(a,\theta) = \sin\theta,$ (ii) $u(a,\theta) = \sin\theta, 0 \leqslant \theta \leqslant \pi$
   $$= 0,\ \pi \leqslant \theta \leqslant 2\pi.$$

3. Solve the equation of exercise 1 under the conditions $u(0,y) = 0$, $\partial u/\partial x = 0$, $x = a$, $0 < y < b$; $u(x,0) = 0$, $u(x,b) = 1$, $0 < x < a$.

4. Solve the boundary value problem of exercise 1 when u is prescribed to be unity on $x = 0$, $x = a$ and $y = 0$.

5. Solve $\partial^2 u/\partial x^2 = \partial u/\partial t$, $0 < x < l$, $t > 0$, subject to the conditions $u(0, t) = a$, $u(l, t) = b$, $t > 0$; $u(x, 0) = 0$, $0 < x < l$; a and b being arbitrary constants.

6. Solve $\partial^2 u/\partial t^2 = \partial^2 u/\partial x^2$, $0 < x < l$, $t > 0$, subject to the conditions $u(0, t) = 0$, $\partial u/\partial x = 1$, $x = l$, $t > 0$; $u(x, 0) = 0$, $\partial u/\partial t = 0$, $t = 0$, $0 < x < l$.

7. Solve the previous exercise when the condition on $x = l$ is $u = \sin t$ [$n\pi \neq l$ for any integer n].

Chapter 3

Fourier Integrals

3.1 The General Fourier Integral   If a function $f(x)$ satisfies the conditions of Theorem 1 for all intervals then (by setting $a = -c$ in the expressions of 1.1) a Fourier series representation for $f(x)$ can be obtained in $[-c, c]$ for all values of c. This series will not, however large the value chosen for c, represent $f(x)$ for all x unless it has period $2c$. We therefore examine the possibility of extending the ideas of §1.1 to represent a function for all x in terms of trigonometric functions.

In the following analysis it is convenient to use the exponential form of the Fourier series described immediately prior to the statement of Theorem 1. Any function $f(x)$ satisfying the conditions of Theorem 1 is such that, except at a discontinuity,

$$f(x) = \sum_{n=-\infty}^{\infty} \alpha_n \exp(in\pi x/c), \qquad |x| \leqslant c,$$

$$\alpha_n = \frac{1}{2c} \int_{-c}^{c} f(x)\exp(-in\pi x/c)\, dx$$

$$= \frac{h}{2\pi} \int_{-c}^{c} f(x)\exp(-inhx)\, dx, \qquad \text{where } ch = \pi.$$

Hence $f(x) = \dfrac{1}{2\pi} \displaystyle\sum_{n=-\infty}^{\infty} hF^{(c)}(nh)\exp(inhx)$

where $F^{(c)}(\beta) = \displaystyle\int_{-c}^{c} f(x)\exp(-i\beta x)\, dx.$

The ordinary definition of an integral as a limit of a sum now suggests that, on taking the limit $h \to 0$,

$$f(x) = \frac{1}{2\pi} \int_{-\infty}^{\infty} \exp(i\beta x)F(\beta)\, d\beta \tag{3.1}$$

where $F(\beta) = \displaystyle\lim_{c \to \infty} F^{(c)}(\beta) = \int_{-\infty}^{\infty} f(x)\exp(-i\beta x)\, dx.$ $\qquad$ (3.2)

$F(\beta)$ is known as the Fourier transform of f.

The above analysis is clearly not rigorous and assumes existence of various limits and integrals; equations (3.1) and (3.2) are, however, valid under fairly general conditions, and their validity can be formalized in the following theorem.

Theorem 8   If $f(x)$ is piecewise smooth on every finite interval of the x-axis and $\int_{-\infty}^{\infty} |f(x)|\, dx$ exists, then the right hand side of equation (3.1) represents

42

(i) $f(x)$ at all points at which it is continuous,

(ii) $\frac{1}{2}[f(x_r+0)+f(x_r-0)]$ at any point of discontinuity $x = x_r$.

An important result associated with Fourier integrals is the Riemann–Lebesgue lemma which states that, if f satisfies the conditions of Theorem 8, $\lim\limits_{\beta \to \infty} F(\beta) = 0$.

The Fourier integral is of such general applicability that tables of functions and their corresponding transforms have been compiled; some relevant results are tabulated in Appendix 2. Thus only very few examples of direct calculation of Fourier transforms will be given. The process of determining $f(x)$ given $F(\beta)$ is known as the inversion of a transform and this process is sometimes helped by the following theorem.

Theorem 9 (*Convolution Theorem*)   If f and g are functions satisfying the conditions of Theorem 8 and $F(\beta)$ and $G(\beta)$ denote their Fourier transforms then the inverse of FG is $\int_{-\infty}^{\infty} f(t)g(x-t)\,dt$.

Problem 3.1   Find the Fourier transform of the function $f(x)$: $f(x) = 0$, $x < 0$; $f(x) = \exp(-x)$, $x > 0$; $f(0) = \frac{1}{2}$. Hence obtain an integral representation of $f(x)$ for all x.

Solution. From equation (3.2) it follows that, since $f \ldots 0$ for $x < 0$,

$$F(\beta) = \int_0^{\infty} \exp(-x-i\beta x)\,dx = \frac{1}{1+i\beta} = \frac{1-i\beta}{1+\beta^2}.$$

$f(x)$ satisfies the conditions of Theorem 8, and hence

$$f(x) = \frac{1}{2\pi}\int_{-\infty}^{\infty} \exp(i\beta x)F(\beta)\,d\beta = \frac{1}{2\pi}\int_{-\infty}^{\infty} \frac{(1-i\beta)}{1+\beta^2}\exp(i\beta x)\,d\beta$$

$$= \frac{1}{2\pi}\left[\int_0^{\infty} \frac{(1-i\beta)\exp(i\beta x)\,d\beta}{1+\beta^2} + \int_{-\infty}^{0} \frac{(1-i\beta)\exp(i\beta x)}{1+\beta^2}\,d\beta\right].$$

β can be replaced by $-\beta$ in the second integral thus:

$$f(x) = \frac{1}{\pi}\int_0^{\infty} \frac{[\cos \beta x + \beta \sin \beta x]}{1+\beta^2}\,d\beta. \qquad \square$$

Problem 3.2   Find the Fourier transform of $\exp(-a|x|)$ $(a > 0)$ and hence obtain an integral representation for $\exp(-a|x|)$.

Solution.

$$F(\beta) = \int_{-\infty}^{\infty} \exp(-a|x|-i\beta x)\,dx$$

$$= \int_0^{\infty} \exp(-ax-i\beta x)\,dx + \int_{-\infty}^{0} \exp(ax-i\beta x)\,dx$$

43

D

$$= \frac{1}{a+i\beta} + \frac{1}{a-i\beta} = \frac{2a}{a^2+\beta^2}.$$

Thus, from Theorem 8,

$$\exp(-a|x|) = \frac{a}{\pi} \int_{-\infty}^{\infty} \frac{\exp(i\beta x)\, d\beta}{a^2+\beta^2}. \qquad \square$$

The above identity is not very easy to obtain by elementary means, but one method would be by means of contour integration and use of Cauchy's theorem and readers familiar with complex variable theory might be interested in this alternative approach. $[\exp(ixz)/a^2+z^2]$ is integrated over the closed contour in the z plane consisting of the real axis from $z = -R$ to $z = R$ and the semicircle in the upper half plane joining these points. There is a simple pole at $z = ai$ with residue $-i\exp(-ax)/2a$, and Jordan's lemma shows that the integral over the semicircle tends to zero as $R \to \infty$, thus applying Cauchy's theorem and taking the limit as $R \to \infty$ gives the result for $x > 0$. The corresponding calculation for $x < 0$ requires the contour to be in the lower half plane; the case $x = 0$ is elementary.

Problem 3.3 If $f(x)$ and $f'(x)$ are continuous functions of x satisfying all the conditions of Theorem 8, show that the Fourier transform of f' is $i\beta F(\beta)$, where F is the Fourier transform of f.

Solution. The required Fourier transform is

$$\int_{-\infty}^{\infty} f'(x)\exp(-i\beta x)\, dx = [f(x)\exp(-i\beta x)]_{x=0}^{x=\infty}$$

$$+ i\beta \int_{-\infty}^{\infty} f(x)\exp(-i\beta x)\, dx, \qquad [\text{on integration by parts}].$$

Since the integral of $|f(x)|$ from $x = -\infty$ to $x = \infty$ converges it is necessary that $f \to 0$ as $x \to \pm\infty$; thus the integrated terms vanish and the required result is proved. $\qquad \square$

Problem 3.4 Show that an even function $f(x)$ satisfying the condition of Theorem 8 has a Fourier representation of the form $\int_0^\infty G(\beta)\cos\beta x\, d\beta$.

Solution. The Fourier transform of $f(x)$ is

$$F(\beta) = \int_{-\infty}^{\infty} f(x)\exp(-i\beta x)\, dx =$$

$$= \int_0^\infty f(x)\exp(-i\beta x)\, dx + \int_{-\infty}^0 f(x)\exp(-i\beta x)\, dx.$$

x can be replaced in the second integral by $-x$ and the evenness property then gives $F(\beta) = 2\int_0^\infty f(x)\cos\beta x\, dx$. Cos βx is an even function of β hence the same is true of F; a particular example of this general result has been derived in Problem 3.2. $f(x)$ is such that Theorem 8 can be applied and thus

$$f(x) = \frac{1}{2\pi} \int_{-\infty}^{\infty} F(\beta)\exp(i\beta x)\, d\beta$$

$$= \frac{1}{2\pi} \int_{0}^{\infty} F(\beta)\exp(i\beta x)\, d\beta + \frac{1}{2\pi} \int_{-\infty}^{0} F(\beta)\exp(i\beta x)\, d\beta.$$

Replacing β by $-\beta$ in the second integral and using the evenness property of $F(\beta)$ gives $f(x) = (1/\pi) \int_0^{\infty} F(\beta)\cos \beta x\, d\beta$, which is of the required form. $\qquad\qquad\square$

It can similarly be shown that, for f an odd function of x,

$$f(x) = (1/\pi) \int_{0}^{\infty} G(\beta)\sin \beta x\, d\beta \quad \text{where} \quad G(\beta) = 2 \int_{0}^{\infty} f(x)\sin \beta x\, dx.$$

3.2 Fourier Cosine and Sine Transforms

By constructing even and odd extensions of a function $f(x)$ defined for $x > 0$, one can (from Problem 3.4) obtain integral representations involving only sines or cosines [this is analogous to obtaining sine and cosine series representations on a finite interval]. Hence we have the following theorem.

Theorem 10 For a function $f(x)$ satisfying the conditions of Theorem 8 for $x > 0$ the Fourier cosine and sine transforms $F_c(\beta)$ and $F_s(\beta)$ defined by

$$F_c(\beta) = \int_{0}^{\infty} f(x)\cos \beta x\, dx, \tag{3.3}$$

$$F_s(\beta) = \int_{0}^{\infty} f(x)\sin \beta x\, dx \tag{3.4}$$

are such that at all points $x > 0$ at which f is continuous,

$$f(x) = (2/\pi) \int_{0}^{\infty} F_c(\beta)\cos \beta x\, d\beta, \tag{3.5}$$

$$f(x) = (2/\pi) \int_{0}^{\infty} F_s(\beta)\sin \beta x\, d\beta. \tag{3.6}$$

At those points $x = x_r$ at which f is discontinuous the left hand sides of equation (3.5) and (3.6) have to be replaced by $\frac{1}{2}[f(x_r+0)+f(x_r-0)]$. At $x = 0$ the right-hand side of equation (3.5) converges to $f(0+0)$ and that of equation (3.6) to zero.

Problem 3.5 Calculate the Fourier cosine and sine transforms of $\exp(-x)\cos x$.

Solution.

$F_c(\beta) = \int_0^{\infty} \exp(-x)\cos x \cos \beta x\, dx, \quad F_s(\beta) = \int_0^{\infty} \exp(-x)\cos x \sin \beta x\, dx.$

The most direct method of calculating these integrals is by expressing

45

$\cos \beta x$ and $\sin \beta x$ in terms of $\exp(\pm i\beta x)$ and taking $\cos x$ as the real part of $\exp(ix)$. Thus, on defining $G(\beta)$ by

$$G(\beta) = \int_0^\infty \exp(-x+ix+i\beta x)\,dx = \frac{1}{1-i(\beta+1)} = \frac{1+i(\beta+1)}{2+2\beta+\beta^2},$$

we have that

$$F_c(\beta) = \tfrac{1}{2}\operatorname{Re}[G(\beta)+G(-\beta)], \qquad F_s(\beta) = \tfrac{1}{2}\operatorname{Re} -i[G(\beta)-G(-\beta)].$$

Hence

$$F_c(\beta) = \frac{1}{2}\frac{\operatorname{Re} 2(1+i)(2+\beta^2)-4i\beta^2}{(2+\beta^2)^2-4\beta^2} = \frac{\beta^2+2}{\beta^4+4},$$

$$F_s(\beta) = \frac{1}{2}\frac{\operatorname{Re} 4i(1+i)\beta+2\beta(2+\beta^2)}{\beta^4+4} = \frac{\beta^3}{\beta^4+4}. \qquad \square$$

From these expressions and Theorem 10 are obtained the identities

$$\exp(-x)\cos x = \frac{2}{\pi}\int_0^\infty \frac{(\beta^2+2)}{\beta^4+4}\cos \beta x\,d\beta$$

$$\exp(-x)\sin x = \frac{2}{\pi}\int_0^\infty \frac{\beta^3 \sin \beta x}{\beta^4+4}\,d\beta.$$

Problem 3.6   Solve the integral equation

$$\int_0^\infty f(\beta)\cos \beta x\,d\beta = \begin{array}{ll} 1, & 0 \leqslant x \leqslant 1, \\ 0, & x > 1. \end{array}$$

Solution. Theorem 10 effectively states that if

$$\int_0^\infty f(\beta)\cos \beta x\,d\beta = F(x)$$

then

$$f(\beta) = (2/\pi)\int_0^\infty F(x)\cos \beta x\,dx,$$

and applying this result gives

$$f(\beta) = \frac{2}{\pi}\int_0^1 \cos \beta x\,dx = \frac{2\sin \beta}{\pi\beta}.$$

Substituting this into the original equation and setting $x = 0$ shows that

$$\int_0^\infty \frac{\sin \beta}{\beta}\,d\beta = \tfrac{1}{2}\pi. \qquad \square$$

Problem 3.7   Find the Fourier cosine transform of $\exp(-x^2)$.

Solution. We have $F_c(\beta) = \int_0^\infty \cos \beta x \exp(-x^2)\,dx$. The simplest method of evaluating this integral is by a complex variable method but we shall consider an alternative method which uses more elementary ideas. The integral is uniformly convergent for $x \geqslant 0$ and thus $dF_c/d\beta$ can be evaluated by differentiating under the integral sign, giving

$$dF_c/d\beta = -\int_0^\infty x \sin \beta x \exp(-x^2) \, dx = \tfrac{1}{2} \int_0^\infty \sin \beta x d \exp(-x^2)$$
$$= -\tfrac{1}{2}\beta F_c,$$

(on integration by parts).

Hence $F_c = A \exp(-\tfrac{1}{4}\beta^2)$ where A is a constant. A may be found by setting $\beta = 0$ and using the known result $2\int_0^\infty \exp(-x^2) \, dx = \sqrt{\pi}$. The final result is $2F_c(\beta) = \sqrt{\pi} \exp(-\tfrac{1}{4}\beta^2)$. $\qquad\square$

[For readers familiar with complex variable theory the following direct method might be of interest. As e^{-x^2} is an even function of x it follows, by an analysis exactly similar to that of Problem 3.4, that

$$2F_c(\beta) = \int_{-\infty}^\infty \exp(-i\beta x - x^2) \, dx = \exp(-\tfrac{1}{4}\beta^2) \int_{-\infty}^\infty \exp(-x + \tfrac{1}{2}i\beta)^2 \, dx.$$

$\exp(-z^2)$ is an analytic function of z and thus, integrating it round the rectangular region with vertices at $z = \pm R, \pm R + \tfrac{1}{2}i\beta$, gives

$$\int_{-R}^R \exp(-z^2) \, dz + \int_R^{R+\tfrac{1}{2}i\beta} \exp(-z^2) \, dz + \int_{R+\tfrac{1}{2}i\beta}^{-R+\tfrac{1}{2}i\beta} \exp(-z^2) \, dz +$$
$$+ \int_{-R+\tfrac{1}{2}i\beta}^{-R} \exp(-z^2) \, dz = 0.$$

As $R \to \infty$ the second and fourth integrals tend to zero, and setting $z = x + \tfrac{1}{2}i\beta$ in the third integral gives

$$\int_{-\infty}^\infty \exp(-z^2) \, dz = \int_{-\infty}^\infty \exp[-(x + \tfrac{1}{2}i\beta)^2] \, dx.$$

The left-hand side of this identity is known to be $\sqrt{\pi}$ and thus $F_c(\beta)$ can be calculated.] $\qquad\square$

Problem 3.8 If f, f' and f'' are continuous functions of x satisfying all the conditions of Theorem 10 determine the cosine and sine transforms of f'' in terms of the corresponding transforms of f and the values of f and f' at $x = 0$.

Solution. The cosine transform of f'' is $G_c(\beta)$ defined by

$$G_c(\beta) = \int_0^\infty f'' \cos \beta x \, dx$$

and integrating by parts we get

$$G_c(\beta) = [f' \cos \beta x]_{x=0}^{x=\infty} + \beta \int_0^\infty f' \sin \beta x \, dx$$
$$= [f' \cos \beta x]_{x=0}^{x=\infty} + \beta [f \sin \beta x]_{x=0}^{x=\infty} - \beta^2 \int_0^\infty f \cos \beta x \, dx$$

(on a further integration by parts). f, f' both vanish as $x \to \infty$ and thus

47

$G_c(\beta) = -[f']_{x=0} - \beta^2 F_c(\beta)$. Repeated integration by parts of the corresponding integral for $G_s(\beta)$, the sine transform of f, gives

$$G_s(\beta) = \beta[f]_{x=0} - \beta^2 F_s(\beta). \qquad \square$$

3.3 Solution of Boundary-Value Problems Using Fourier Transforms

Fourier cosine and sine transforms are useful for the solution of boundary value problems for

$$a_1 \frac{\partial^2 u}{\partial x^2} + a_3 u + b_1(y) \frac{\partial^2 u}{\partial y^2} + b_2(y) \frac{\partial u}{\partial y} + b_3(y)u = f(x, y), \ x > 0, 0 < y < d,$$

where a_1, a_3 are constant, with $u \to 0$ as $x \to \infty$ and either u or $\partial u/\partial x$ prescribed on $x = 0$. The general method is to take either the sine or cosine transform of the above equation; this gives (using the results of Problem 3.8)

$$a_1[-\beta^2 U_c - (\partial u/\partial x)_{x=0}] + a_3 U_c + \int_0^\infty Lu \cos \beta x \, dx = F_c, \qquad (3.7)$$

$$a_1[-\beta^2 U_s + \beta(u)_{x=0}] + a_3 U_s + \int_0^\infty Lu \sin \beta x \, dx = F_s, \qquad (3.8)$$

where capitals denote Fourier transforms, the suffixes c and s have the same meaning as before, and L denotes the operator $b_1 \partial^2/\partial y^2 + b_2 \partial/\partial y + b_3$. In deriving the above results it has been assumed that u, $\partial u/\partial x$, $\partial^2 u/\partial x^2$ are continuous in x and satisfy the conditions of Theorem 8. It is also assumed that differentiation with respect to y may be interchanged with the integration operation in equations (3.7) and (3.8) giving

$$a_1[-\beta^2 U_c - (\partial u/\partial x)_{x=0}] + a_3 U_c + LU_c = F_c, \qquad (3.9)$$

$$a_1[-\beta^2 U_s + \beta(u)_{x=0}] + a_3 U_s + LU_s = F_s. \qquad (3.10)$$

Thus, if $\partial u/\partial x$ is known at $x = 0$, equation (3.9) becomes a differential equation for U_c regarded as a function of y; similarly, if u is prescribed at $x = 0$, equation (3.10) gives the differential equation satisfied by U_s. Once either U_c or U_s has been found then Theorem 10 can be used to find u. Various assumptions have had to be made about the unknown function and it has to be verified that the function eventually determined satisfies these conditions. The above general procedure is formally identical with that described in §2.3 for the solution of inhomogeneous boundary value problems using Fourier series.

If x ranges from $-\infty$ to ∞ and the conditions on u are $u \to 0$ as $x \to \pm\infty$, then the exponential Fourier transform has to be used. In this case it is possible to consider the more general equation

$$a_1 \frac{\partial^2 u}{\partial x^2} + a_2 \frac{\partial u}{\partial x} + a_3 u + Lu = f, \qquad (3.11)$$

48

where a_1, a_2, a_3 are constant. Taking the Fourier transform of equation (3.11) gives, using the result of Problem 3.3,

$$-\beta^2 a_1 U + i\beta a_2 U + a_3 U + LU = F. \tag{3.12}$$

Various assumptions concerning continuity and interchanging operators have again been made in deriving equation (3.12). This is a differential equation for U and, after it has been solved, u may be found from Theorem 8.

Problem 3.9 Solve $\partial^2 u/\partial x^2 = \partial u/\partial t$, $-\infty < x < \infty$, $t > 0$ under the conditions $u \to 0$ as $|x| \to \infty$, $u(x,0) = \exp(-x^2)$.

Solution. The appropriate transform to use is clearly the exponential one, and taking the transform of the equation gives $-\beta^2(U) = \partial U/\partial t$. At $t = 0$, $u = \exp(-x^2)$ and hence $U = \int_{-\infty}^{\infty} \exp(-i\beta x - x^2)\,dx$ at $t = 0$, and it follows from Appendix 2, equation (5) that $U = \pi^{\frac{1}{2}}\exp(-\frac{1}{4}\beta^2)$ when $t = 0$. Solving the differential equation for U gives

$$U = A(\beta)\exp(-\beta^2 t)$$

where A can be a function of β. The condition on $t = 0$ shows that $A(\beta)$ is equal to the value of U at $t = 0$ and hence

$$U = \pi^{\frac{1}{2}}\exp[-\tfrac{1}{4}\beta^2(1+4t)].$$

The inverse of $\pi^{\frac{1}{2}}\exp(-\frac{1}{4}\beta^2)$ is $\exp(-x^2)$ [Appendix 2, (5)] and from Appendix 2, (1) it follows that the inverse of $\exp(-\frac{1}{4}\lambda^2\beta^2)$ is $\lambda^{-1}\exp(-\lambda^{-2}x^2)$. Hence setting $\lambda = (1+4t)^{\frac{1}{2}}$ shows that $u = (1+4t)^{-\frac{1}{2}}\exp(-x^2)/(1+4t)$. It can be verified in this case, by direct substitution, that u is the required solution. $\square$

Problem 3.10 Solve $\partial^2 u/\partial x^2 + \partial^2 u/\partial y^2 = 0$, $0 < x < \infty$, $0 < y < d$, under the conditions $u(x,0) = u(x,d) = 0$, $x > 0$; $u(0,y) = 1 - y/d$, $0 < y < d$; $u \to 0$ as $x \to \infty$.

Solution. This problem could be solved by either representing u as a Fourier sine series in y or taking the Fourier sine transform with respect to x, and we choose the latter approach. Taking the Fourier sine transform of the above equation gives

$$\frac{\partial^2 U_s}{\partial y^2} - \beta^2 U_s = -\beta(1 - y/d), \qquad \text{[c.f. equation (3.10)]}.$$

The general solution of this equation is

$$U_s = A\cosh \beta y + B\sinh \beta y + (d-y)/\beta d,$$

and the conditions $U_s = 0$ for $y = 0$ and $y = d$ [which follow from $u = 0$ on $y = 0, d$] yield

$$U_s = \frac{\sinh \beta(y-d)}{\beta \sinh \beta d} + \frac{(d-y)}{\beta d}.$$

Hence $u = \dfrac{2}{\pi} \displaystyle\int_0^\infty \dfrac{\sinh \beta(y-d)\sin \beta x d\beta}{\beta \sinh \beta d} + \dfrac{2}{\pi}\dfrac{(d-y)}{d}\displaystyle\int_0^\infty \dfrac{\sin \beta x d\beta}{\beta}.$

Setting $\alpha = y-d$, $\gamma = d$ in Appendix 2, equation (8) gives

$$\int_0^\infty \frac{\sinh x(y-d)\sin \beta x}{x \sinh dx}\,dx = \tan^{-1}\left\{\tan\left[\frac{\pi}{2d}(y-d)\right]\tanh\left(\frac{\pi\beta}{2d}\right)\right\}$$

and interchanging the variables β and x gives the first term in u. The second term in u can be evaluated by using the identity derived at the end of Problem 3.6 (replacing β by βx) and thus finally

$$u = 1 - \frac{y}{d} - \frac{2}{\pi}\tan^{-1}\left[\cot\frac{\pi y}{2d}\tanh\frac{\pi x}{2d}\right].$$

Clearly this function satisfies all the conditions imposed. □

Problem 3.11 Solve the equation of the previous problem in $0 < y < d$, $x > 0$, subject to the conditions $u(x,0) = 0$, $u(x,d) = 1$; $x > 0$; $u(0, y) = 0, 0 < y < d$.

Solution. The sine transform is still applicable and now

$$\frac{\partial^2 U_s}{\partial y^2} - \beta^2 U_s = 0.$$

As u vanishes on $y = 0$, U_s must also vanish on $y = 0$; $u(x, d) = 1$ and hence U_s on $y = d$ is the sine transform of unity. However, unity does not satisfy the conditions of Theorem 10 and it does not seem possible to calculate its Fourier sine transform. It follows however from the last identity of Problem 3.6 by replacing β by βx, $(x > 0)$, that there exists an $F(\beta)$ (in this case $2/\pi\beta$) such that $\int_0^\infty [F(\beta)\sin \beta x]d\beta = 1$. $F(\beta)$ is not absolutely integrable and thus Theorem 10 cannot be used to invert the above result. We therefore seek an integral representation for u of the form $u = \int_0^\infty G(\beta, y)\sin \beta x\,d\beta$. Substituting this in the equation for u shows that G satisfies the same equation as U_s above. The boundary conditions give $G(\beta, d) = 2/\pi\beta$ and $G(\beta, o) = 0$, and hence $G = 2\sinh \beta y/\beta \sinh \beta d$ and

$$u = \frac{2}{\pi}\int_0^\infty \frac{\sinh \beta y \sin \beta x\,d\beta}{\beta \sinh \beta d}.$$

This integral is of the same form as that of Problem 3.10 with $y-d$ now replaced by y and hence

50

$$u = \frac{2}{\pi} \tan^{-1}\left[\tan\frac{\pi y}{2d} \tanh\frac{\pi x}{2d}\right]. \qquad \square$$

(For all problems with homogeneous boundary conditions the above method of assuming an appropriate integral representation of the solution can be used instead of taking the transform of the equation. The two methods are entirely equivalent but, provided the relevant transforms exist, taking the transform of the equation has the advantage of also being applicable for non-homogeneous boundary conditions. This is also the method used for solving ordinary differential equations using Laplace transforms.)

Problem 3.12 Solve the equation of Problem 3.9 for $x > 0$, $t > 0$ subject to the conditions $u \to 0$ as $x \to \infty$, $u(x, 0) = 0$ and $\partial u/\partial x = 1$, $x = 0, t > 0$.

Solution. In this case the only possibility is the cosine transform and $-\beta^2 U_c - 1 = \partial U_c/\partial t$ [c.f. equation (3.9)]. The general solution is

$$U_c = -\frac{1}{\beta^2} + A(\beta)\exp - \beta^2 t$$

and, as $U_c = 0$ for $t = 0$, it is necessary that $A = \beta^{-2}$. Thus applying Theorem 10 gives

$$u = \frac{2}{\pi} \int_0^\infty \frac{\cos \beta x(\exp(-\beta^2 t) - 1)}{\beta^2} \, d\beta.$$

Appendix 2, equation (12) gives (interchanging x and β)

$$\int_0^\infty \exp(-\alpha\beta^2)\cos \beta x d\beta = \tfrac{1}{2}(\pi/\alpha)^{\frac{1}{2}}\exp(-x^2/4\alpha)$$

and integrating both sides of this with respect to α from $\alpha = 0$ to $\alpha = t$ gives

$$\int_0^\infty \frac{1 - \exp(-\beta^2 t)}{\beta^2}\cos \beta x \, d\beta = \tfrac{1}{2}\pi^{\frac{1}{2}} \int_0^t \alpha^{-\frac{1}{2}}\exp(-x^2/4\alpha) \, d\alpha.$$

Making the change of variable $v^2 = x^2/4\alpha$ enables the integral on the right-hand side to be rewritten as

$$-x \int_\infty^{x/2t} \exp(-v^2)/v^2 \, dv = 2t^{\frac{1}{2}}\exp(-x^2/4t) + 2x \int_\infty^{\infty/2t} \exp(-v^2)dv,$$

on integration by parts.

Hence $\qquad u = -2t^{\frac{1}{2}}\exp(-x^2/4t)/\pi^{\frac{1}{2}} + x\,\mathrm{Erfc}(x/2t^{\frac{1}{2}}),$

where Erfc z, the complementary error function, is defined by
Erfc $z = 2\pi^{-\frac{1}{2}} \int_z^\infty \exp(-t^2)dt$. $\qquad\qquad\qquad\qquad \square$

Problem 3.13 Solve

$$\frac{\partial^2 u}{\partial r^2} + \frac{1}{r}\frac{\partial u}{\partial r} + \frac{\partial^2 u}{\partial z^2} = 0, \qquad 0 < r < a, z > 0$$

under the conditions $u(0, z)$ finite, $u(a, z) = 0$ and $u(r, 0) = 1$.

Solution. This problem can be solved by taking the sine transform with respect to z and this gives

$$\frac{\partial^2 U_s}{\partial r^2} + \frac{1}{r}\frac{\partial U_s}{\partial r} - \beta^2 U_s = -\beta.$$

A particular integral of the equation is $1/\beta$ and the corresponding homogeneous equation is Bessel's equation of order zero in the variable $i\beta r$. The appropriate finite solution of this is $J_0(i\beta r)$ which is normally written as $I_0(\beta r)$.

The solution satisfying the condition on $r = a$ is thus

$$U_s = \frac{1}{\beta}\left[1 - \frac{I_0(\beta r)}{I_0(\beta a)}\right]$$

Hence
$$u = \frac{2}{\pi}\int_0^\infty \left[1 - \frac{I_0(\beta r)}{I_0(\beta a)}\right]\frac{\sin \beta z}{\beta}\, d\beta.$$

The first term is equal to unity (c.f. last identity of Problem 3.6) and the second integral can be transformed by use of Cauchy's theorem to yield the form of solution obtained by using a Fourier–Bessel expansion (cf. Chapter 5). From the numerical point of view the infinite integral form is as convenient as the infinite series form. □

Problem 3.14 Solve $\partial^2 u/\partial x^2 + \partial^2 u/\partial y^2 = 0$, $y > 0$, all x, under the conditions $u(x, 0) = f(x)$, $u \to 0$ as $x^2 + y^2 \to \infty$, where f is a function satisfying the conditions of Theorem 8.

Solution. The two possible methods of solution are the exponential Fourier transform with respect to x or a sine transform with respect to y. The latter will lead to an inhomogeneous equation for the Fourier transform and this generally presents more technical difficulties. We therefore take the exponential Fourier transform with respect to x and have $\partial^2 U/\partial y^2 - \beta^2 U = 0$. In order to satisfy the conditions at infinity $U = \exp(-|\beta|y)$ and to satisfy the condition on $y = 0$, $A(\beta)$ must be equal to $F(\beta)$, the Fourier transform of f. Hence $U = F(\beta)\exp(-|\beta|y)$.

Theorem 8 now gives a formal expression for u but a simpler one is obtained by applying Theorem 9 (the convolution theorem) and using the result that the inverse of $\exp(-|\beta|y)$ is $y/\pi(x^2 + y^2)$ [Appendix 2, (4)].

Hence

$$u = \frac{y}{\pi} \int_{-\infty}^{\infty} \frac{f(\xi)\, d\xi}{(x-\xi)^2 + y^2}.$$ □

Problem 3.15 Solve

$$\frac{\partial^4 u}{\partial x^4} + 2\frac{\partial^4 u}{\partial x^2 \partial y^2} + \frac{\partial^4 u}{\partial y^4} = 0, \qquad y > 0, \text{ all } x,$$

under the conditions $\partial u/\partial y = 0$, $y = 0$; $(\partial u/\partial x)_{y=0} = f(x)$, where $f(x)$ is a function satisfying the conditions of Theorem 8. Also $u \to 0$ as x, $y \to \infty$.

Solution. This equation is not of the general type that we have considered, but by analogy with our previous examples it would seem reasonable to attempt a Fourier transform and use of the exponential form avoids any complications with inhomogeneous terms. Thus

$$\beta^4 U - 2\beta^2 \frac{\partial^2 U}{\partial y^2} + \frac{\partial^4 U}{\partial y^4} = 0,$$

the general form of solution tending to zero as $y \to \infty$ is $U = [A + By]\exp(-|\beta|y)$, the condition on $\partial u/\partial y$ on $y = 0$ gives $|\beta|A = B$. The condition on $\partial u/\partial x$ on $y = 0$ gives $i\beta U = F$, where F is the Fourier transform of f and thus

$$U - \frac{i}{\beta}(1 + |\beta|y)\exp(-|\beta|y)F(\beta).$$

The inverse of the term multiplying $F(\beta)$ is

$$-\frac{i}{2\pi} \int_{-\infty}^{\infty} \frac{(1 + |\beta|y)}{\beta} \exp(-|\beta|y)\exp(i\beta x)\, d\beta$$

$$= \frac{1}{\pi} \int_0^{\infty} \frac{\sin \beta x}{\beta} \exp(-\beta y)\, d\beta + \frac{y}{\pi} \int_0^{\infty} \sin \beta x \exp(-\beta y)\, d\beta.$$

From Appendix 2(9) the first term of the above equation is equal to $(1/\pi)\tan^{-1}(x/y)$ and Appendix 2(12) shows that the second integral is equal to $x/\pi(x^2 + y^2)$. The solution for u can now be written down by means of theorem 9. □

Problem 3.16 Solve Problem 3.12 when the condition on $x = 0$ is $\partial u/\partial x = hu$ where h is a constant and $u(x, 0) = \exp(-x)$.

Solution. If h were zero then the appropriate transform would be a Fourier cosine transform and for $h \to \infty$ the Fourier sine transform would be appropriate but neither is suitable for other values of h. It

53

therefore seems worth considering the possibility of using a combination of sine and cosine transforms and we consider U^* defined by

$$U^* = a \int_0^\infty u \sin \beta x \, dx + b \int_0^\infty u \cos \beta x \, dx,$$

where a and b are functions of β. To obtain the differential equation satisfied by U^* it is necessary to calculate

$$\int_0^\infty (a \sin \beta x + b \cos \beta x) \frac{\partial^2 u}{\partial x^2} \, dx = \left[(a \sin \beta x + b \cos \beta x) \frac{\partial u}{\partial x} \right]_{x=0}^{x=\infty}$$

$$- \int_0^\infty (\beta a \cos \beta x - \beta b \sin \beta x) \frac{\partial u}{\partial x} \, dx$$

$$= \left[(a \sin \beta x + b \cos \beta x) \frac{\partial u}{\partial x} - (\beta a \cos \beta x - \beta b \sin \beta x) u \right]_{x=0}^{x=\infty}$$

$$- \beta^2 \int_0^\infty (a \sin \beta x + b \cos \beta x) u \, dx.$$

For any of the integrals to converge it is necessary that $u \to 0$ as $x \to \infty$; this means the integrated terms vanish as $x \to \infty$ and

$$\int_0^\infty (a \sin \beta x + b \cos \beta x) \frac{\partial^2 u}{\partial x^2} \, dx = \left[-b \frac{\partial u}{\partial x} + a \beta u \right]_{x=0} - \beta^2 U^*,$$

and, in view of the boundary condition on u at $x = 0$, the first term vanishes if $a\beta = hb$ and the simplest choice of a and b gives $a = h$; $b = \beta$. Multiplying the equation for u by $h \sin \beta x + \beta \cos \beta x$ and integrating with respect to x from 0 to ∞ formally gives $\partial U^*/\partial t = -\beta^2 U^*$.

At $t = 0$, U is $\exp(-x)$ and hence

$$U^* = \int_0^\infty (h \sin \beta x + \beta \cos \beta x) \exp(-x) \, dx = \beta(h+1)/(1+\beta^2),$$

Thus $U^* = (1+h)\beta \exp(-\beta^2 t)/(1+\beta^2)$. It is now necessary to determine the inversion formula for this transform, we have

$$U^* = \int_0^\infty hu \sin \beta x \, dx + \int_0^\infty u \frac{d}{dx}(\sin \beta x) \, dx = \int_0^\infty \left(hu - \frac{\partial u}{\partial x} \right) \sin \beta x \, dx,$$

(on integrating the second term by parts). From this expression it is seen that U^* is the sine transform of $hu - \partial u/\partial x$ and hence

$$hu - \frac{\partial u}{\partial x} = \frac{2}{\pi} \int_0^\infty U^* \sin \beta x \, d\beta.$$

This equation can be rewritten as

$$-\frac{\partial}{\partial x}(u e^{-hx}) = \frac{2}{\pi} e^{-hx} \int_0^\infty U^* \sin \beta x \, d\beta$$

and replacing x by a variable w and integrating both sides with respect to w from $w = x$ to $w = \infty$ gives

$$u = \frac{2}{\pi} \int_0^\infty \frac{(h \sin \beta x + \beta \cos \beta x) U^*}{h^2 + \beta^2} \, d\beta.$$

The above derivation of the inversion formula has made various assumptions concerning interchanging orders of integration, but it can be verified that the inversion formula is valid for all functions $f(x)$ satisfying the conditions of Theorem 8. Thus the solution of the present problem can be obtained in the form of an infinite integral. $\quad\square$

EXERCISES

1. If $f(x)$ is $\exp(2x)$ for $x \leqslant 0$ and zero for $x > 0$ show that

$$f(x) = \frac{1}{\pi} \int_0^\infty \frac{2 \cos \beta x - \beta \sin \beta x}{\beta^2 + 4} \, d\beta.$$

2. Obtain the Fourier sine transform of $\exp(-x)\sin x$, $(x > 0)$ and hence show that, for integer $n, \int_0^\infty (x \sin n\pi x)/(x^4 + 4) \, dx = 0$.

3. Solve the integral equation

$$= 0, \quad 0 \leqslant \beta \leqslant 1,$$
$$\int_0^\infty f(x)\sin \beta x \, dx = 1, \quad 1 < \beta \leqslant 2,$$
$$= 0, \quad \beta > 2.$$

4. Solve $\partial u/\partial t = \partial^2 u/\partial x^2$, $x, t > 0$, subject to the conditions $u(0, t) = \sin \omega t, u \to 0$ as $x \to \infty, t > 0; u(x, 0) = 0, x > 0$.

5. Solve $\partial^2 u/\partial x^2 - 3u = \partial u/\partial t$, $x, t > 0$ under the conditions $u(x, 0) = 0$, $x > 0; \partial u/\partial x = e^{-t}, x = 0, t > 0; \lim_{x \to \infty} u(x, t) = 0, t > 0$.

Chapter 4

Generalized Series Expansions

4.1 Introduction   In Chapter 1 it was shown that a wide class of functions could be represented by series of trigonometric functions and we now examine the extension of this principle to represent a function as an infinite series of certain other types of functions. In practice it is only useful to consider a representation of f on $[a, b]$ by a series of the form $\sum_{n=1}^{\infty} a_n \phi_n$ where the ϕ_n satisfy a relation of the form

$$\int_a^b \rho \phi_n \phi_m \, dx = 0, \qquad m \neq n, \tag{4.1}$$

where ρ is a given function. In this case the functions ϕ_n are said to be **orthogonal** with weight ρ on $[a, b]$. All the Fourier series previously considered are of this form with $\rho = 1$ (c.f. Problem 1.3).

If it is assumed that, for a given set of functions ϕ_n, a function f is such that

$$f = \sum_{n=1}^{\infty} a_n \phi_n, \qquad a < x < b, \tag{4.2}$$

then, multiplying both sides of this equation by $\rho \phi_m$ and integrating with respect to x from $x = a$ to $x = b$ gives

$$\int_a^b \rho f \phi_m \, dx = \sum_{n=1}^{\infty} a_n \int_a^b \rho \phi_n \phi_m \, dx,$$

(assuming that the order of summation and integration may be interchanged). Equation (4.1) now shows that

$$\int_a^b \rho f \phi_m \, dx = a_m \int_a^b \rho \phi_m^2 \, dx, \tag{4.3}$$

thus yielding an expression for a_m. The above analysis is purely formal as it need not be true that, for a given set of ϕ_n, equation (4.2) should hold, for any f. (If the ϕ_n were chosen to be $1, \cos 2x, \cos 3x, \ldots$ on $[0, \pi]$ then for $f = \cos x$, equation (4.3) would give $a_m \equiv 0$, thus yielding an obvious contradiction and showing that the above set of functions would not be suitable to use in representing a function on $[0, \pi]$.) In those cases where equation (4.2) is meaningful the coefficients defined by equation (4.3) are known as the generalised Fourier coefficients of f and the resulting series representation as a generalized Fourier representation of f. The formal procedure given in Problem 1.3 is a special case of the method outlined above.

56

The above example shows that a necessary condition for a representation of the form of the right-hand side of equation (4.3) to be valid is that there exists no function orthogonal to all the ϕ_n. In making statements of this nature concerning the existence of a function, one has to be somewhat more precise and specify the class of functions considered (e.g. functions continuous on an interval, or the class of integrable or square integrable functions over an interval). The class of functions is sometimes said to constitute a function space. If the functions ϕ_n are orthogonal with weight ρ on $[a, b]$ and such that there is no function f of the same class, orthogonal to all the ϕ_n and such that $\int_a^b \rho f^2 \, dx \neq 0$, then the set of functions ϕ_n is said to be **complete** within the given class. If the series of equation (4.2) with a_m defined by equation (4.3), converges to f, except at possibly a finite number of points in $[a, b]$, for all functions f of the class considered, then the set ϕ_n is said to be **closed** in the sense of pointwise convergence. This terminology is introduced here to make the reader aware of it but the concepts will not be used in the text.

We now examine briefly two general classes of problems which lead naturally to the idea of expanding a function in terms of two particular sets of functions. We consider first the problem of solving Laplace's equation in a spherical region when axi-symmetric boundary conditions are prescribed on one, or two, spherical surfaces. Laplace's equation becomes, on assuming axial symmetry,

$$\frac{1}{r^2} \frac{\partial}{\partial r} \left(r^2 \frac{\partial u}{\partial r} \right) + \frac{1}{r^2 \sin \theta} \frac{\partial}{\partial \theta} \left(\sin \theta \frac{\partial u}{\partial \theta} \right) = 0,$$

where r, θ are the usual spherical polar coordinates. Writing u as $F_1(r)F_2(\theta)$ and using the procedure of separation of variables gives

$$\frac{1}{r^2} \frac{d}{dr} \left(r^2 \frac{dF_1}{dr} \right) - \lambda F_1 = 0,$$

$$\frac{1}{\sin \theta \, d\theta} \left(\sin \theta \frac{dF_2}{d\theta} \right) + \lambda F_2 = 0, \tag{4.4}$$

and, on writing x as $\cos \theta$, equation (4.4) becomes

$$\frac{d}{dx} \left[(1 - x^2) \frac{dF_2}{dx} \right] + \lambda F_2 = 0. \tag{4.5}$$

In physical problems F_2 has to be finite for all θ (i.e. $|x| \leqslant 1$) and it can be shown from the theory of ordinary differential equations that equation (4.5) only has such finite solutions when $\lambda = n(n+1)$, where n is an integer.

57

In this case there is a finite solution, the Legendre polynomial, $P_n(x)$, of degree n satisfying

$$\frac{d}{dx}\left[(1-x^2)\frac{dP_n}{dx}\right] + n(n+1)P_n = 0, \qquad (4.6)$$

and the condition $P_n(1) = 1$.

When $\lambda = n(n+1)$ independent solutions for F_1 are r^n and r^{-n-1}, hence an appropriate series representation for u is $\sum\limits_{n=0}^{\infty} (A_n r^n + B_r r^{-n-1})P_n(\cos\theta)$. Thus if u is prescribed for $r = a$ and $r = b$ in terms of θ and if any function of θ can be expressed as a series of Legendre polynomials then A_n and B_n can be found. Boundary value problems of this type lead to the concept of expanding a function as a series of Legendre polynomials. The formal properties of such an expansion are given in a subsequent section but it is convenient to summarise some of the simpler properties of these functions at this stage.

An explicit form for $P_n(x)$ is provided by Rodrigues' formula

$$P_n(x) = \frac{1}{2^n n!}\frac{d^n}{dx^n}(x^2-1)^n, \qquad (4.7)$$

some particular cases of this being

$$P_0(x) = 1, \qquad (4.8)$$

$$P_1(x) = x, \qquad (4.9)$$

$$2P_2(x) = 3x^2 - 1. \qquad (4.10)$$

Also

$$P_n(-x) = (-1)^n P_n(x), \qquad (4.11)$$

$$P_{2n}(0) = \frac{1.3\ldots 2n-1}{2.4\ldots 2n}, \quad n = 1, 2, \ldots, \qquad (4.12)$$

$$P_{2n+1}(0) = 0, \qquad (4.13)$$

$$P'_{2n}(0) = 0, \qquad (4.14)$$

$$P'_{2n+1}(0) = (2n+1)P_{2n}(0), \qquad (4.15)$$

$$P_0(0) = P'_1(0) = 1, \qquad (4.16)$$

$$(1-2xt+t^2)^{-\frac{1}{2}} = \sum_{n=0}^{\infty} t^n P_n(x). \qquad (4.17)$$

The Legendre polynomials are orthogonal with unit weight on $[-1, 1]$ and

$$\int_{-1}^{1} P_n^2(x)\, dx = \frac{2}{(2n+1)}. \qquad (4.18)$$

Laplace's equation in cylindrical polar coordinates (ρ, ϕ, z) becomes, assuming axial symmetry,

$$\frac{1}{\rho}\frac{\partial}{\partial \rho}\left(\rho\frac{\partial u}{\partial \rho}\right)+\frac{\partial^2 u}{\partial z^2} = 0,$$

writing $u = G_1(\rho)G_2(z)$ and using separation of variables gives

$$\frac{1}{\rho}\frac{d}{d\rho}\left(\rho\frac{dG_1}{d\rho}\right)+\lambda G_1 = 0, \qquad (4.19)$$

$$\frac{d^2 G_2}{dz^2}-\lambda G_2 = 0.$$

Equation (4.19) can be rewritten as

$$\frac{d^2 G_1}{dy^2}+\frac{1}{y}\frac{dG_1}{dy}+G_1 = 0,$$

where $y = \lambda^{\frac{1}{2}}\rho$; this is Bessel's equation of order zero with solutions $J_0(y)$ and $Y_0(y)$. For finite solutions in boundary value problems for regions including the origin Y_0 must be excluded. For the particular class of problem when u is required to vanish on $\rho = 1$ it is necessary that $J_0(\lambda^{\frac{1}{2}}) = 0$ and thus $\lambda^{\frac{1}{2}}$ must be one of the zeros of $J_0(y)$. There are an infinite number of these which we shall denote by j_{0n}. The equation for G_2 is soluble in simple terms and thus the appropriate series representation is $\sum_{n=0}^{\infty} (A_n \cosh j_{0n} z + B_n \sinh j_{0n} z)J_0(j_{0n}\rho)$. If u is prescribed on $z = a$ and $z = b$ and if any function of ρ can be expanded as a series of $J_0(j_{0n}\rho)$ then A_n and B_n may be found.

In §3.3 the general concept of expanding a function in $[0, 1]$ as a series of terms of the form $J_v(j_{vn}x)$, where $J_v(j_{vn}) = 0$, is examined in detail. Some of the simpler properties of Bessel functions are summarised below.

The Bessel function $J_v(\lambda x)$ satisfies the equation

$$\frac{d^2 J_v}{dx^2}+\frac{1}{x}\frac{dJ_v}{dx}+\left(\lambda^2-\frac{v^2}{x^2}\right)J_v = 0, \qquad (4.20)$$

and

$$J_v(\lambda x) = \sum_{r=0}^{\infty}\frac{(-1)^r(\frac{1}{2}x)^{v+2r}}{r!(v+r)!}. \qquad (4.21)$$

Also

$$d[x^v J_v(\lambda x)]/dx = \lambda x^v J_{v-1}(\lambda x), \qquad (4.22)$$

$$d[x^{-v} J_v(\lambda x)]/dx = -\lambda x^{-v} J_{v+1}(\lambda x), \qquad (4.23)$$

$$J_{v-1}(x)+J_{v+1}(x) = \frac{2v}{x}J_v(x), \qquad (4.24)$$

$$J_{v-1}(x)-J_{v+1}(x) = 2J_v'(x), \qquad (4.25)$$

where the dash denotes the derivative with respect to the argument. The

E

set of functions $J_v(j_{vn} x)$ $[n = 1, 2, \dots]$ are orthogonal with weight x on $[0, 1]$ and

$$\int_0^1 x J_v^2(j_{vn} x) \, dx = \tfrac{1}{2} J_{v+1}^2(j_n).\tag{4.26}$$

Series involving $J_v(j_{vn} x)$ are generally known as Fourier–Bessel series.

4.2 Expansion in Terms of Legendre Functions The basic result is given by the following theorem.

Theorem 11 If f is piecewise smooth in $[-1, 1]$ then the series $\sum\limits_{n=0}^{\infty} a_n P_n(x)$, where $2a_n = (2n + 1) \int_{-1}^{1} f(x) P_n(x) \, dx$, converges for $|x| < 1$ to $f(x)$ at all points at which f is continuous and to $\tfrac{1}{2}[f(x+0) + f(x-0)]$ at points of discontinuity. At the end points $x = +1$, $x = -1$ the series converges to $f(1-0)$ and $f(-1+0)$ respectively. The particular form occurring for a_n can be deduced directly from equations (4.3) and (4.18).

$$\int_{-1}^{1} f(x) P_n(x) \, dx = \frac{1}{2^n n!} \int_{-1}^{1} f(x) \frac{d^n}{dx^n} (x^2 - 1)^n \, dx, \quad \text{[from equation 4.7]}$$

$$= \frac{1}{2^n n!} \left[f(x) \frac{d^{n-1}}{dx^{n-1}} (x^2 - 1)^n \right]_{-1}^{1} - \frac{1}{2^n n!} \int_{-1}^{1} \frac{df}{dx} \frac{d^{n-1}}{dx^{n-1}}$$

$$(x^2 - 1)^n \, dx, \text{[on integration by parts.]}$$

$(x^2 - 1)^n$ and its first $(n-1)$ derivatives vanish at $x = 1$, $x = -1$ and thus the integrated part in the above expression vanishes. The process can be repeated n times giving

$$\int_{-1}^{1} f(x) P_n(x) \, dx = \frac{(-1)^n}{2^n n!} \int_{-1}^{1} \frac{d^n f}{dx^n} (x^2 - 1)^n \, dx.\tag{4.27}$$

Equation (4.27) is often useful in calculating coefficients in Legendre expansions.

Problem 4.1 Find the Legendre expansion on $[-1, 1]$ of x^2.

Solution. Equation (4.27) and Theorem 11 show that the coefficient of $P_n(x)$ is given by

$$a_n = \frac{(-1)^n (2n+1)}{2^{n+1} n!} \int_{-1}^{1} \frac{d^n}{dx^n} (x^2)(x^2 - 1)^n \, dx.$$

The nth derivative of x^2 is zero for $n > 2$ hence $a_n = 0$, $n > 2$ and

$$a_2 = \tfrac{5}{8} \int_{-1}^{1} (x^2 - 1)^2 \, dx = \tfrac{2}{3}, \qquad a_1 = -\tfrac{3}{2} \int_{-1}^{1} x(x^2 - 1) \, dx = 0,$$

$$a_0 = \tfrac{1}{2} \int_{-1}^{1} x^2 \, dx = \tfrac{1}{3}.$$

Hence $x^2 = \frac{2}{3}P_2 + \frac{1}{3}P_0$; this particular result could have been obtained more directly from equations (4.8) and (4.10). ☐

It can be shown similarly that the coefficients of P_n in the expansion of x^m vanish for $n > m$. In this case the Legendre expansion would only have a finite number of terms and is best obtained by comparing the coefficients of powers of x in the expansion $x^m = \sum_{n=0}^{m} a_n P_n(x)$, where the explicit expressions for the Legendre polynomials are substituted on the right-hand side.

Problem 4.2 Expand $\cos 2\phi$ as a series in $P_n(\cos \phi)$ in $0 \leqslant \phi \leqslant \pi$.

Solution. In this example x has been replaced by $\cos \phi$ and the same substitution should be made in the integral for a_n. It is, however, simpler in this case to use $\cos 2\phi = 2\cos^2\phi - 1$ and equations (4.8) and (4.10). Thus

$$\cos 2\phi = 2\cos^2\phi - 1 = \frac{2}{3}[2P_2(\cos \phi) + 1] - 1$$
$$= \frac{4}{3}P_2(\cos \phi) - \frac{1}{3} = \frac{4}{3}P_2(\cos \phi) - \frac{1}{3}P_0(\cos \phi) \quad (\text{since } P_0 = 1). \quad ☐$$

Problem 4.3 Find the Legendre expansion in $[-1, 1]$ of $f(x)$:

$$f(x) = 0, \quad (-1 \leqslant x < 0), \quad f(x) = 1, \quad (0 \leqslant x \leqslant 1).$$

Solution. In this case $a_n = \frac{1}{2}(2n+1)\int_0^1 P_n(x)\, dx$. The integral can be evaluated by using equation (4.7) but a more direct approach is to use equation (4.6), giving

$$a_n = -\frac{(2n+1)}{2n(n+1)} \int_0^1 \frac{d}{dx}\left[(1-x^2)\frac{dP_n}{dx}\right] dx = \frac{2n+1}{2n(n+1)} P_n'(0).$$

$P_n'(0)$ can be calculated from equations (4.13) to (4.16), which show that it is zero for n even ($n \neq 0$) and

$$a_{2k+1} = \frac{(-1)^k 4k + 3}{(4k+2)(2k+2)} \frac{1.3.5....(2k-1)}{2.4.6...2k} \quad k = 1, 2, ...$$
$$a_1 = \frac{3}{4}, \qquad a_0 = \frac{1}{2}.$$

By theorem 11 the series should converge to $\frac{1}{2}$ at $x = 0$, which is a point of discontinuity and the above results confirm this as, from equation (4.13), $P_{2n+1}(0) = 0$ and the value of the series at $x = 0$ is $a_0 = \frac{1}{2}$. ☐

Problem 4.4 Show that it is possible to represent a function $f(x)$ defined in $[0, 1]$ as a series of Legendre polynomials $P_n(x)$ of even order.

Solution. Since Theorem 11 refers to the interval $[-1, 1]$ it is necessary

61

to find a function $F(x)$ equal to $f(x)$ in $[0, 1]$ and such that the coefficients of odd order in the Legendre expansion of F are zero, i.e.

$$\int_{-1}^{1} F(x)P_n(x)\, dx = \int_{0}^{1} P_n(x)F(x)\, dx + \int_{-1}^{0} P_n(x)F(x)\, dx$$

$$= \int_{0}^{1} P_n(x)[F(x)+(-1)^n F(-x)] \times dx = 0, \qquad n \text{ odd}$$

on replacing x by $-x$ in the second integral and using equation (4.11). Thus $F(-x) = F(x)$ and hence $F(x)$ has to be defined for $-1 \leqslant x < 0$ as $f(-x)$. The coefficients a_{2n} of $P_{2n}(x)$ are given by

$$a_{2n} = \frac{4n+1}{2} \int_{-1}^{1} F(x)P_{2n}(x)\, dx = 4n+1 \int_{0}^{1} f(x)P_{2n}(x)\, dx.$$

The series converges at $x = 0$ to $\frac{1}{2}[F(0+0)+F(0-0)] = f(0)$. $\qquad \square$

Similarly $f(x)$ can be represented as a series of Legendre polynomials of odd order by defining $F(x) = -f(-x)$ for $-1 \leqslant x < 0$.

In this case the resulting series converges at $x = 0$ to $\frac{1}{2}[F(0+0)+F(0-0)] = 0$.

Problem 4.5 Find the series representing x on $[0, 1]$ in terms of Legendre polynomials of even order.

Solution. This is a special case of the previous example with $f(x) = x$ and hence $a_{2n} = (4n+1) \int_{0}^{1} x\, P_{2n}(x)\, dx$. The integral is best calculated by using equation (4.7) giving

$$a_{2n} = \frac{4n+1}{2^{2n}(2n)!} \int_{0}^{1} x \frac{d^{2n}}{dx^{2n}}(x^2-1)^{2n}\, dx$$

$$= \frac{(4n+1)}{2^{2n}(2n)!} \left\{ \left[x \frac{d^{2n-1}(x^2-1)^{2n}}{dx^{2n-1}} \right]_{x=0}^{x=1} - \int_{0}^{1} \frac{d^{2n-1}(x^2-1)^{2n}}{dx^{2n-1}}\, dx \right\},$$

$$n \neq 0.$$

The integrated part vanishes at both limits and thus, after a further integration,

$$a_{2n} = -\frac{(4n+1)}{2^{2n}(2n)!} \left[\frac{d^{2n-2}}{dx^{2n-2}}(x^2-1)^{2n} \right]_{x=0}^{x=1}, \qquad n \neq 0,$$

the expression in square brackets vanishes at $x = 1$ and can be evaluated at $x = 0$ by expanding $(x^2-1)^{2n}$ by means of the binomial theorem, yielding

$$a_{2n} = \frac{4n+1}{2^{2n}(2n)!} \frac{d^{2n-2}}{dx^{2n-2}} \sum_{r=0}^{2n} \frac{x^{2r}(-1)^{2n-r}(2n)!}{r!(2n-r)!}, \qquad x = 0.$$

It is only the term x^{2n-2} that produces a non-zero contribution after differentiating $2n-2$ times and setting $x = 0$, and thus

$$a_{2n} = \frac{(4n+1)(2n-2)!(-1)^{n+1}}{2^{2n}(n-1)!(n+1)!}, \qquad n \neq 0.$$

a_0 can be obtained directly from the initial expression (since $P_0 = 1$) and is equal to $\frac{1}{2}$. $\qquad\qquad\qquad\qquad\qquad\qquad\qquad\qquad\qquad\qquad\quad\square$

Problem 4.6 Expand $(5-4x)^{-\frac{1}{2}}$ as a series of Legendre polynomials.

Solution. The straightforward approach using Theorem 11 will lead to evaluating integrals of the form $\int_{-1}^{1}(5-4x)^{-\frac{1}{2}}P_n(x)\,dx$ and this does not seem to be a very simple task. It is, however, worth checking in the case of an expression of the form $(a+bx)^{-\frac{1}{2}}$ whether it is a particular case of the left-hand side of equation (4.17). In the present example

$$(5-4x)^{-\frac{1}{2}} = (4-4x+1)^{-\frac{1}{2}} = \tfrac{1}{2}(1-2\tfrac{1}{2}x+\tfrac{1}{4})^{-\frac{1}{2}}.$$

This suggests applying equation (4.17) with $t = \frac{1}{2}$, giving

$$(5-4x)^{-\frac{1}{2}} = \tfrac{1}{2} \sum_{n=0}^{\infty} 2^{-n}P_n(x). \qquad\qquad\qquad\qquad\quad\square$$

4.3 Fourier–Bessel Series In the present section we examine the possibility of expanding a function on $[0,1]$ as a series of terms proportional to $J_\nu(j_{\nu n}x)$ where $J_\nu(j_{\nu n}) = 0$, and the relevant expansion theorem is:

Theorem 12 If $f(x)$ is piecewise smooth in $[0,1]$ then the series

$\sum_{n=1}^{\infty} a_n J_\nu(j_{\nu n}x)$, where

$$a_n = \frac{2}{J_{\nu+1}^2(j_{\nu n})} \int_0^1 xf(x)J_\nu(j_{\nu n}x)dx,$$

converges to $f(x)$ at all points at which f is continuous and to $\frac{1}{2}[f(x+0)+f(x-0)]$ at points of discontinuity. The series converges to zero at $x = 1$ for all ν and for $\nu > 0$ converges to zero at $x = 0$. The particular form for a_n may be deduced from equations (4.3) and (4.26).

Problem 4.7 Obtain the representation in $[0,1]$ of unity as a series of functions $J_0(j_n x)$ where $J_0(j_n) = 0$.

Solution. This is a particular case of Theorem 12 with $\nu = 0$ and $f = 1$ and thus

$$a_n = \frac{2}{J_1^2(j_n)} \int_0^1 xJ_0(j_n x)\,dx.$$

63

The integral can be evaluated by applying equation (4.22) with $v = 0$ and $\lambda = j_n$, giving

$$a_n = \frac{2}{j_n J_1^2(j_n)} \left[x J_1(j_n x) \right]_{x=0}^{x=1} = \frac{2}{j_n J_1(j_n)}. \qquad \square$$

Problem 4.8 If μ_n are the roots of $J_0(4\mu) = 0$, obtain an expansion in a series of $J_0(\mu_n x)$ of the function $f(x)$ defined on $[0,4]$ by $f(x) = 1$, $0 < x < 2$, $f(x) = 0$, $2 < x \leqslant 4$, $f(2) = \frac{1}{2}$.

Solution. Defining a new variable y by $\frac{1}{4}x$ gives an expansion problem in $[0,1]$ in the new variable. The j_{vn} of Theorem 12 have now to be replaced by $4\mu_n$ and thus the coefficient a_n of $J_0(j_{0n} y)[= J_0(\mu_n x)]$ is

$$\frac{2}{J_1^2(4\mu_n)} \int_0^1 y f(4y) J_0(4\mu_n y) \, dy = \frac{2}{J_1^2(4\mu_n)} \int_0^{\frac{1}{2}} y J_0(4\mu_n y) \, dy = \frac{J_1(2\mu_n)}{4\mu_n J_1^2(4\mu_n)}$$

(equation (4.22), with $v = 1$, $\lambda = 4\mu_n$). $\qquad \square$

The only cases where the coefficients can be evaluated are those when the integrals occurring are such that the equations (4.22) or (4.23) can be used. There are some other special cases where the coefficients take on a comparatively simple form, but these cases require a detailed knowledge of the properties of Bessel functions. Further expansion problems for Fourier–Bessel series and their generalisations occur in Problems 5·2, 5·4, 5·5, 5·6, 5·9 and 5·10.

4.4 Sturm–Liouville Theory It is effectively shown in Problem 2.1 that the set of functions $\sin n\pi x / c (n = 1, 2, \ldots)$ is generated by determining λ such that $X'' = -\lambda X$ has non-trivial solutions in $[0, c]$ vanishing at the end points of the interval. This type of problem is known as an eigen value problem, the appropriate values of λ are the eigen values and the corresponding solutions are known as the eigen functions. Another example is provided in Problem 2.3, where it is shown that the functions $\sin (2n-1)\pi x / 2l$ are eigen functions for the above equation in $[0, l]$ when the conditions at the end points are $X(0) = X'(l) = 0$; the corresponding eigen values being $(2n-1)^2 \pi^2 / 4l^2$.

Both the above examples are special cases of a general class of eigen value problem, the Sturm–Liouville problem, which is such that an arbitrary function can be represented in terms of the solutions to the eigen value problem.

The Sturm–Liouville problem in its most general form is the determination of the eigen values and the corresponding eigen functions in $[a, b]$ of the differential equation:

$$\frac{d}{dx}\left(p\frac{d\phi}{dx}\right)+q\phi+\lambda\rho\phi = 0, \tag{4.28}$$

where p, q, ρ are given functions and the boundary conditions at the end points are either

$$\alpha_1\phi+\alpha_2\frac{d\phi}{dx} = 0, \qquad x = a, \tag{4.29}$$

$$\beta_1\phi+\beta_2\frac{d\phi}{dx} = 0, \qquad x = b, \tag{4.30}$$

or

$$k_1\phi(a)+k_2\phi(b) = 0 = m_1\left(\frac{d\phi}{dx}\right)_{x=a}+m_2\left(\frac{d\phi}{dx}\right)_{x=b}. \tag{4.31}$$

The first class of boundary conditions are known as unmixed boundary conditions and the second class are normally called periodic boundary conditions. If $\rho \neq 0$ in $[a,b]$ then the problem is said to be regular.

The principal results of Sturm–Liouville theory may be summarized in the following theorem.

Theorem 13 Let p, p', $q\rho$ and $(\rho p)''$ be continuous real functions of x on $[a, b]$ with $p > 0$, $\rho > 0$ and the constants α_1, α_2, β_1, β_2, k_1, k_2, m_1, m_2 be real. Then the Sturm–Liouville problem has an infinite number of eigen values λ_n all real and not more than a finite number being negative. The eigen functions ϕ_n, ϕ_m corresponding to different eigen values λ_n, λ_m are orthogonal with weight ρ.

Any continuous function $f(x)$, with a piecewise continuous first derivative, can be represented in $a < x < b$ as an infinite series of the form of equation (4.2) with the coefficients defined by equation (4.3). Furthermore if the function f satisfies the same boundary conditions as those of the Sturm–Liouville problem then the series converges to f in $[a, b]$.

For the case of unmixed boundary conditions (i.e. equations (4.29), (4.30)) there cannot be two linear independent eigen functions corresponding to the same eigen value. Also if the additional conditions $\alpha_1\alpha_2 \leqslant 0$, $\beta_2\beta_1 \geqslant 0$, are imposed then $\lambda_n \geqslant 0$.

The orthogonality conditions also hold for the unmixed problem with $p(a) = 0$ provided that the equation (4.29) is dropped and with $p(b) = 0$ provided that equation (4.30) is dropped.

Note. The Legendre and Fourier–Bessel expansions of the previous sections are particular examples of irregular Sturm–Liouville problems and the orthogonality of these sets of functions is a direct consequence of the above statement concerning orthogonality for the particular case when

$p(a)$ and /or $p(b)$ are zero. The boundary conditions for the Sturm–Liouville problem for the Legendre expansion are that ϕ is bounded at $x = 1$, -1 and those for the Fourier–Bessel expansion are that ϕ is bounded at $x = 0$ and vanishes at $x = 1$.

The set $\sin n\pi x/c$ clearly corresponds to the case $p = 1 = \rho$, $q = 0$ in $[0,c]$ with $\alpha_2 = \beta_2 = 0$ and the sets $\cos n\pi x/c$, $\sin(2n-1)\pi x/c$, $\cos(2n-1)\pi x/2c$, are also generated by the same equation with different sets of values of α, β [for $\cos n\pi x/c$, $\alpha_1 = \beta_1 = 0$; for $\sin(2n-1)\pi x/2c$, $\alpha_2 = \beta_1 = 0$; for $\cos(2n-1)\pi x/2c$, $\alpha_1 = \beta_2 = 0$]. The various expansions quoted in Chapter 1 are particular cases of Theorem 13.

Problem 4.9 Find the Sturm–Liouville problem in $[0, 2c]$ whose eigen functions are $\cos n\pi x/c$, $\sin n\pi x/c$.

Solution. These functions satisfy $X'' + \lambda X = 0$ with $\lambda = n^2\pi^2/c^2$, $\cos n\pi x/c$, $\sin n\pi x/c$ do not satisfy the same mixed conditions at $x = 0$ and $x = c$ but satisfy the periodic conditions $\phi(0) = \phi(2c)$, $\phi'(0) = \phi'(2c)$, and we thus examine the Sturm–Liouville problem for the above equation with these conditions. For $\lambda \neq 0$ the general solution is $A \cos \lambda^{\frac{1}{2}}x + B \sin \lambda^{\frac{1}{2}}x$ and the boundary conditions give

$$A = A \cos 2c\lambda^{\frac{1}{2}} + B \sin 2c\lambda^{\frac{1}{2}}, \qquad B = B \cos 2c\lambda^{\frac{1}{2}} - A \sin 2c\lambda^{\frac{1}{2}}.$$

These equations will only have non-zero solutions if their determinant vanishes and this gives $\cos 2c\lambda^{\frac{1}{2}} = 1$ and hence $\lambda = n^2\pi^2/c^2$. If this value of λ is substituted into the equations they degenerate, and hence any values of A and B are possible and in particular $\cos n\pi x/c$, $\sin n\pi x/c$ are both eigen functions, illustrating that the result concerning the non-multiplicity of eigen functions is not valid for periodic conditions.

For $\lambda = 0$ the general solution is $Cx + D$ and the conditions give $D = 2Cc + D$, $C = C$, and hence $C = 0$ and in this case there is only one eigen function, i.e. $\cos n\pi x/c$ $(n = 0)$. $\qquad \square$

Problem 4.10 Determine the eigen functions and eigen values for $d^2\phi/dx^2 + \lambda\phi = 0$, $0 < x < 1$, with $\phi(0) = 0$ and $h\phi(1) + \phi'(1) = 0$.

Solution. For $\lambda \neq 0$ the general solution is $A \cos \lambda^{\frac{1}{2}}x + B \sin \lambda^{\frac{1}{2}}x$ and as $\phi(0) = 0$ we have that $A = 0$; the condition at $x = 1$ gives $hB \sin \lambda^{\frac{1}{2}} + \lambda^{\frac{1}{2}}B \cos \lambda^{\frac{1}{2}} = 0$. In this case the product $\beta_1 \beta_2$ of Theorem 13 is equal to h and hence, for $h > 0$, λ will be positive thus $\lambda^{\frac{1}{2}}$ will be real. This can also be deduced directly by writing $\lambda = -u^2$ which gives $hB \sinh u + Bu \cosh u = 0$, and, if B is not identically zero, $\tanh u = -u/h$ and for $h > 0$ this is not possible. If h is negative then, as $\tanh u$ is a monotonic increasing function of u, there may be one solution. The slope of the tanh curve, however, is less than unity and, as both sides of the equation are equal for $u = 0$, it thus follows that the curves will not intersect for $h > -1$. Thus for $h < -1$ there will be one negative eigen value $\lambda = -u^2$, where u is the positive root of $\tanh u = -u/h$ and the corresponding eigen function is proportional to $\sinh ux$.

The remaining eigen values are given by $\lambda_n = v_n^2$ where the v_n are the positive roots of $h \tan v = -v$. (It is obvious from sketching the graph of $\tan v$ that there will be an infinite number of solutions of this equation; this is also implied by Theorem 13.) The eigen functions are proportional to $\sin v_n x$.

The general solution for $\lambda = 0$ is $\phi = Cx + D$, and since ϕ vanishes at $x = 0$ it follows that D must vanish and the condition at $x = 1$ gives $Ch + C = 0$. Thus for the case $h = -1$, $\lambda = 0$ will be an eigen value with the corresponding eigen function being proportional to x. $\qquad\square$

Problem 4.11 Determine the eigen values and eigen functions for $x^2 d^2\phi/dx^2 + 3xd\phi/dx + \lambda\phi = 0$, in $[1, e]$ with ϕ vanishing at the end points of the interval.

Solution. This is not in Sturm–Liouville form but can be rewritten in that form with $p = x^3$, $\rho = x$. The first step is the determination of the general solution and the form of the equation suggests considering ϕ to be of the form x^μ and this gives $(\mu+1)^2 + \lambda - 1 = 0$. Thus for $\lambda \neq 1$ the general solution is

$$\phi = Ax^{(1-\lambda)^{\frac{1}{2}}-1} + Bx^{-(1-\lambda)^{\frac{1}{2}}-1}.$$

The condition at $x = 1$ gives $A = -B$ and application of the condition at $x = e$ gives $A \sinh(1-\lambda)^{\frac{1}{2}} = 0$. Thus $(1-\lambda)^{\frac{1}{2}} = in\pi$ and $\lambda = 1 + n^2\pi^2$, $n = 1, 2, \ldots$, and the eigen functions are proportional to $x^{-1}[x^{in\pi} - x^{-in\pi}]$.

The above analysis does not show that $\lambda = 1$ is not an eigen value and this case has to be treated separately. For $\lambda = 1$ it can be verified that independent solutions of the equation are x^{-1} and $x^{-1}\ln x$, and thus the general solution is $x^{-1}[C + D\ln x]$.

The condition at $x = 1$ gives $C = 0$, and from $\phi(e) = 1$ it follows that $D \equiv 0$. Hence $\lambda = 1$ is not an eigen value. $\qquad\square$

Problem 4.12 Obtain, for the case $h > 0$, an expansion of unity in $[0, 1]$ in terms of the eigen functions of Problem 4.10.

Solution. The eigen functions are proportional to $\sin v_n x$ where $h \tan v_n = -v_n$ and thus, by Theorem 13 and equation (4.3),

$$1 = \sum_{n=1}^{\infty} a_n \sin v_n x,$$

where $\int_0^1 \sin v_n x\, dx = a_n \int_0^1 \sin^2 v_n x\, dx = \frac{1}{2} a_n \int_0^1 (1 - \cos 2v_n x)\, dx$, hence $a_n = 4(1 - \cos v_n)/(2 - \sin 2v_n)$. $\qquad\square$

In using equation (4.3) we have assumed the result that the eigen functions of a Sturm–Liouville problem are orthogonal, and in this simple case we shall give a direct verification of this general result. Integrating by parts, we have

$$\int_0^1 \sin v_n x \sin v_m x\, dx = \frac{1}{v_n} \sin v_m \cos v_n + \frac{v_m}{v_n} \int_0^1 \cos v_n x \cos v_m x\, dx$$

$$= \frac{1}{v_n} \sin v_m \cos v_n - \frac{v_m}{v_n^2} \sin v_n \cos v_m$$

$$+ \frac{v_m^2}{v_n^2} \int_0^1 \sin v_n x \sin v_m x\, dx.$$

Hence

$$\left(1 - \frac{v_m^2}{v_n^2}\right) \int_0^1 \sin v_n x \sin v_m x\, dx = \frac{\cos v_m \cos v_n}{v_n}\left[\tan v_m - \frac{v_m}{v_n} \tan v_n\right].$$

As v_n, v_m both satisfy $(\tan v)/v = -1/h$ the right-hand side vanishes and hence, for $v_m \neq v_n$, so does the integral on the left-hand side.

EXERCISES

1. Expand x^5 in $[-1, 1]$ as a series of Legendre polynomials.

2. Expand as a series of Legendre polynomials $f(x)$ defined by:
$$f(x) = 0, \; -1 \leqslant x < \alpha, \; f(x) = 1, \; \alpha \leqslant x \leqslant 1.$$
   [Hint: Use $(2n+1)P_n(x) = P'_{n+1}(x) - P'_{n-1}(x)$.]

3. Obtain the Fourier–Bessel expansion of x^3 on $[0, 1]$.

4. Obtain the equation for the eigen values of $d^2y/dx^2 + \lambda y = 0$, subject to the conditions $y(0) + y'(0) = 0 = 2y(\pi) + 3y'(\pi)$.

Chapter 5

Solution of Boundary-Value Problems by Means of Generalized Fourier Expansions

It is possible to use the results of Chapter 4 to extend the methods of Chapter 2 to problems when the functions a_1, a_2, a_3 are not constant and when a linear combination of u and $\partial u/\partial x$ is prescribed on $x = 0$ and $x = c$. Problems where the equation has the form of equation (2.2) and a linear combination of u and $\partial u/\partial x$ is required to vanish at $x = 0$ and $x = c$ will be termed homogeneous problems and, as in Chapter 2, the cases of homogeneous and non-homogeneous problems will be considered separately.

5.1 Homogeneous Problems The technique is a simple generalization of that of §2.2 A product solution XY is assumed and separation of variables used to determine one of the functions (which will be assumed to be X) as the solution of an eigen value problem and the principle of superposition is then used to obtain a general representation. If the eigen value problem is of a type such that any function can be expanded in terms of the eigen functions then the boundary conditions and equation for the corresponding Y can be obtained as for Fourier coefficients. It is possible to avoid some of the manipulative work of separation of variables in special cases by remembering the general form of product solutions of certain equations. It is shown in §4.1 that the general form of product solution of Laplace's equation finite for all θ in a spherical region is

$$\sum_{n=0}^{\infty} (A_n r^n + B_n r^{-n-1}) P_n(\cos \theta).$$

Another useful result which follows from the analysis of §4.1 is that $[A J_0(\lambda\rho) + B Y_0(\lambda p)][C \cosh \lambda z + D \sinh \lambda z]$, where Y_0 is the second solution of Bessel's equation, satisfies the axisymmetric Laplace equation.

Problem 5.1 Find a solution u of Laplace's equation, valid outside the sphere of radius a, tending to zero at infinity and such that on the sphere $u = \cos \theta$, where θ is the angle between the radius vector and a fixed direction. Obtain also the solution to the problem when the condition on the sphere is $u = (a^2 + f^2 - 2af \cos \theta)^{-\frac{1}{2}}$, $f > a$.

[The first problem corresponds to determining the total electrostatic potential when an earthed sphere is placed in a field of uniform intensity. The second problem arises when the uniform field is replaced by that of a point charge at a distance f from the centre of the sphere.]

Solution. If spherical polar coordinates (r, θ) are chosen with origin at the centre of the sphere, then the analysis of §4.1 shows that an appropriate expansion to choose for u is

$$u = \sum_{n=0}^{\infty} (A_n r^n + B_n r^{-n-1}) P_n(\cos \theta).$$

This form automatically satisfies Laplace's equation and is finite for all real θ. In order to satisfy the condition at infinity $A_n \equiv 0$ and the condition on $r = a$ gives

$$\sum_{n=0}^{\infty} B_n a^{-n-1} P_n(\cos \theta) = \cos \theta.$$

Thus we need to expand $\cos \theta$ as a series of Legendre polynomials, but $\cos \theta = P_1(\cos \theta)$, hence $B_n = 0, n \neq 1, B_1 = a^2$.

To solve the second problem we require the expansion of

$$(a^2 + f^2 - 2af \cos \theta)^{-\frac{1}{2}} = f^{-1}[1 + (a/f)^2 - 2(a/f)\cos \theta]^{-\frac{1}{2}}$$

and equation (4.17) shows that

$$[1 + (a/f)^2 - 2(a/f)\cos \theta]^{-\frac{1}{2}} = \sum_{n=0}^{\infty} (a/f)^n P_n(\cos \theta),$$

hence $B_n = a^{2n+1}/f^{n+1}$. The solution is thus

$$u = \frac{1}{r} \sum_{n=0}^{\infty} \frac{a}{f}\left(\frac{a^2}{fr}\right)^n P_n(\cos \theta) = \frac{a}{f}[r^2 + a^4/f^2 - 2(a^2 r/f)\cos \theta]^{-\frac{1}{2}},$$

from equation (4.17). The quantity in square brackets is the inverse distance from the point (r, θ) to $(a^2/f, 0)$ and the above form of solution is that obtained directly in the electrostatic problem by the method of images. □

Problem 5.2 Solve

$$\frac{\partial^2 u}{\partial r^2} + \frac{1}{r}\frac{\partial u}{\partial r} = \frac{\partial^2 u}{\partial t^2}, \qquad 0 \leqslant r < a, t > 0,$$

subject to the conditions
$u(0, t)$ bounded, $u(a, t) = 0$, $t > 0$, $u(r, 0) = (1 - r^2/a^2)$, $\partial u(r, t)/\partial t = 0$, $t = 0, 0 \leqslant r < a$.
(This corresponds to determining the displacement at any time of a circular membrane of radius a which is fixed round its edge, the initial displacement is $1 - r^2/a^2$ and the membrane is initially at rest.)

Solution. We assume a solution of the form $F(r)G(t)$ and the method of separation of variables gives

70

$$\frac{d^2F}{dr^2}+\frac{1}{r}\frac{dF}{dr}+\lambda^2 F = 0, \qquad \frac{d^2T}{dt^2}+\lambda^2 T = 0.$$

The equation for F is Bessel's equation of order zero, in order that $u(0,t)$ be bounded F must be $J_0(\lambda r)$ and the condition on $r = a$ gives $J_0(\lambda a) = 0$. Thus $\lambda a = j_{0n}$ and to avoid unduly complicated expressions we drop the suffix zero. Thus the appropriate form of solution is

$$u = \sum_{n=0}^{\infty} J_0(j_n r/a)[A_n \cos j_n t/a + B_n \sin j_n t/a].$$

Since $\partial u/\partial t = 0$ for $t = 0$ it follows that $B_n = 0$ and thus A_n is the coefficient of $J_0(j_n r/a)$ in the expansion of $1 - r^2/a^2$ in terms of $J_0(j_n r/a)$ or, equivalently, of $J_0(j_n x)$ in the Fourier–Bessel series in $[0,1]$ for $(1-x^2)$ (this follows on writing r/a as x). Theorem 12 gives

$$A_n = \frac{2}{J_1^2(j_n)} \int_0^1 (1-x^2)x J_0(j_n x)\, dx.$$

The first part of this integral has already been evaluated in Problem 4.7 and the second part can be evaluated as follows:

$$\int_0^1 x^3 J_0(j_n x)\, dx = \frac{1}{j_n}\int_0^1 x^2 \frac{d}{dx}[x J_1(j_n x)]\, dx, \qquad \text{from (4.22), with } v = 1,$$

$$= \frac{J_1(j_n)}{j_n} - \frac{2}{j_n}\int_0^1 x^2 J_1(j_n x)\, dx,$$

$$\text{on integration by parts,}$$

$$= \frac{J_1(j_n)}{j_n} - \frac{2}{j_n^2}\int_0^1 \frac{d}{dx}[x^2 J_2(j_n x)]\, dx, $$

$$\text{from (4.22), with } v = 2,$$

$$= \frac{J_1(j_n)}{j_n} - \frac{2J_2(j_n)}{j_n^2}.$$

Thus from the above result and that of Problem 4.7,

$$A_n = \frac{4J_2(j_n)}{j_n^2 J_1^2(j_n)},$$

this can be further simplified by using equation (4.24) and the fact that $J_0(j_n) = 0$ and the final result is $A_n = 8/j_n^3 J_1(j_n)$. □

Problem 5.3 Solve $\partial u/\partial t = \partial^2 u/\partial x^2$, $0 < x < 1$, $t > 0$, under the conditions $u(0,t) = 0$, $\partial u/\partial x + hu = 0$, $x = 1$, $t > 0$; $u(x,0) = 0$, $0 < x < 1$, h being a positive constant.

Solution. The occurrence of $\partial^2 u/\partial x^2$ suggests a Fourier series in x but the condition on $x = 1$ is not appropriate for Fourier series. We therefore assume product solutions of the form $X(x)T(t)$ and carrying out the normal separation of variables procedure gives

$$X'' + \lambda X = 0, \qquad T' + \lambda T = 0.$$

The conditions on $x = 0$ and $\dot{x} = 1$ give $X(0) = 0$, $X'(1) + hX(1) = 0$ and thus the eigen value problem for X is precisely that of Problem 4.10. Hence, using the notation of that problem,

$$u = \sum_{n=1}^{\infty} A_n \exp(-v_n^2 t)\sin v_n x,$$

and on setting $t = 0$ it follows that the A_n are the generalized Fourier coefficients of x. The weight ρ in this case is unity and thus

$$A_n \int_0^1 \sin^2 v_n x \, dx = \int_0^1 x \sin v_n x \, dx,$$

i.e.
$$v_n(2v_n - \sin 2v_n)A_n = 4(\sin v_n - v_n \cos v_n). \qquad \square$$

Problem 5.4 Solve

$$\frac{\partial^2 u}{\partial r^2} + \frac{1}{r}\frac{\partial u}{\partial r} + \frac{\partial^2 u}{\partial z^2} = 0, \qquad 0 < z < h, 0 < r < a,$$

subject to the conditions $u(0, z)$ finite, $u(r, 0) = 1$, $\partial u(r, z)/\partial r + \gamma u = 0$, $r = a$; $\partial u(r, z)/\partial z + \beta u = 0$, $z = h$, where β, γ are positive constants.

[One physical interpretation of this problem is the determination of the steady state temperature in a cylinder when one of its plane faces is maintained at unit temperature while a radiation condition is imposed on the other surfaces.]

Solution. The application of separation of variables to this equation has been considered in §4.1 and appropriate product solutions satisfying the finiteness condition are $J_0(\lambda r)[A \cosh \lambda(z - h) + B \sinh \lambda(z - h)]$. Since homogeneous conditions are imposed on $z = h$ the algebra is simplified slightly by using hyperbolic functions of $z - h$. The boundary condition on $r = a$ gives

$$\frac{d}{dr}J_0(\lambda r) + \gamma J_0(\lambda r) = 0, \qquad r = a,$$

i.e.
$$\lambda J_0'(\lambda a) + \gamma J_0(\lambda a) = 0,$$

where the prime denotes derivative with respect to the argument. $[J_0'(z) = -J_1(z)]$. Thus λa will have to satisfy the equation $\mu J_0'(\mu) + \delta J_0(\mu) = 0$ where $\delta = \gamma a$. It can be shown that there are an infinite number of solutions, all positive, of this equation and of its generalization $\mu J_\nu'(\mu) + \delta J_\nu(\mu) = 0$.

It can also be shown that any function $f(x)$ continuous in $[0, 1]$ can be represented by a series of the form $\sum_{n=1}^{\infty} a_n J_\nu(\mu_{\nu n} x)$ where $\mu_{\nu n}$ are the roots of the above equation. The coefficients a_n are given by (G. N. Watson *Theory of Bessel Functions*, C.U.P., 1952, Chapter 18)

$$a_n = \frac{2\mu_{\nu n}^2 \int_0^1 t f(t) J_\nu(\mu_{\nu n} t)\, dt}{(\mu_{\nu n}^2 - \nu^2) J_\nu^2(\mu_{\nu n}) + \mu_{\nu n}^2 J_\nu'^2(\mu_{\nu n})}.$$

This more general type of series is known as a Dini series and is an obvious generalization of the Fourier–Bessel series (which corresponds to the case $\delta \to \infty$.). For the particular case $\nu + \delta = 0$ the correct expansion for a function as a Dini series is (G. N. Watson, *Theory of Bessel Functions*, C.U.P., 1952, Chapter 18)

$$\sum_{n=1}^{\infty} a_n J_\nu(\mu_{\nu n} x) + 2(\nu + 1) x^\nu \int_0^1 t^{\nu+1} f(t)\, dt.$$

Reverting to the original boundary value problem it is now seen that the appropriate representation is

$$\sum_{n=1}^{\infty} J_0(\mu_n r/a)[A_n \cosh \mu_n(z-h)/a + B_n \sinh \mu_n(z-h)/a],$$

where the suffix zero is dropped to avoid unnecessarily complicated forms. In order to satisfy the condition at $z = h$, $\mu_n B_n + \beta a A_n = 0$ and the condition on $z = 0$ then gives

$$\sum_{n=1}^{\infty} J_0(\mu_n r/a) A_n[\cosh \mu_n h/a + (\beta a/\mu_n) \sinh \mu_n h/a] = 1,$$

and A_n can now be calculated from the quoted result for Dini series, the integral being evaluated as in Problem 4.7. The final result is

$$A_n[\mu_n \cosh \mu_n h/a + \beta a \sinh \mu_n h/a] = 2J_1(\mu_n)/J_0^2(\mu_n) + J_1^2(\mu_n). \qquad \square$$

Problem 5.5 Solve

$$\frac{\partial^2 u}{\partial r^2} + \frac{1}{r} \frac{\partial u}{\partial r} = \frac{\partial u}{\partial t}, \qquad t > 0, 0 < r < 1,$$

under the conditions $u(0, t)$ finite, $\partial u(r, t)/\partial r = 0$, $r = 1$, $t > 0$; $u(r, 0) = r$, $0 \leqslant r < 1$.

Solution. Following the standard separation of variables procedure shows that there are separable solutions of the form

$$\{A J_0(\lambda r) + B Y_0(\lambda r)\} \exp - \lambda^2 t.$$

The finiteness condition gives $B = 0$ and the condition on $r = 1$ gives

$J_0'(\lambda) = 0$. Thus the set of values for λ are β_n where β_n are the non-zero roots of $J_0'(x) = 0$ and λ can also be zero. The general form for u is

$$a_0 + \sum_{n=1}^{\infty} a_n J_0(\beta_n r)\exp(-\beta_n^2 t)$$

and the condition on $t = 0$ gives

$$a_0 + \sum_{n=1}^{\infty} a_n J_0(\beta_n r) = r.$$

The expansion problem is thus a particular case of the Dini series of the previous problem and it is also one of the special cases when $v + \delta = 0$. Thus

$$a_0 = 2 \int_0^1 t^2 \, dt = \tfrac{2}{3}, \qquad a_n = \frac{2}{J_0^2(\beta_n)} \int_0^1 t^2 J_0(\beta_n t) \, dt. \qquad \square$$

In this case the integral for a_n cannot be simplified.

Problem 5.6   Solve

$$\frac{\partial}{\partial x}\left(x \frac{\partial u}{\partial x}\right) = \frac{\partial^2 u}{\partial t^2}, \qquad 0 < x < l, \quad t > 0,$$

under the conditions $u(0, t)$ bounded, $u(l, t) = 0$, $t > 0$; $u(x, 0) = f(x)$, where f is a given continuous function of x, $\partial u(x, t)/\partial t = 0, t = 0, 0 < x < l$, (This problem arises in the investigation of the small oscillations in the vertical plane of a heavy chain displaced from rest).

Solution. This equation is not one for which the product forms of solution can be instantly recognized, but writing u as $X(x)T(t)$ gives

$$\frac{d}{dx}\left(x \frac{dX}{dx}\right) + \lambda X = 0, \qquad \frac{d^2 T}{dt^2} + \lambda T = 0.$$

X is required to be finite at $x = 0$ and zero at $x = l$ and the problem of determining X is a singular Sturm–Liouville problem. A change to the independent variable $y(= x^{\frac{1}{2}})$ gives

$$\frac{1}{y} \frac{d}{dy}\left(y \frac{dX}{dy}\right) + 4\lambda X = 0,$$

This is Bessel's equation of order zero in the variable $2\lambda^{\frac{1}{2}}y$ and, reverting back to x, the solution finite at $x = 0$ is $J_0(2\lambda^{\frac{1}{2}}x^{\frac{1}{2}})$. The condition at $x = l$ gives $2(\lambda l)^{\frac{1}{2}} = j_n$ where the j_n are defined as in Problem 5.2, and thus the appropriate form of solution is

$$u = \sum_{n=1}^{\infty} \left\{ A_n \cos \frac{j_n t}{2l^{\frac{1}{2}}} + B_n \sin \frac{j_n t}{2l^{\frac{1}{2}}} \right\} J_0[j_n(x/l)^{\frac{1}{2}}].$$

The condition on $\partial u/\partial t$ at $t = 0$ shows that $B_n = 0$ and the other condition at $t = 0$ gives

$$f(x) = \sum_{n=1}^{\infty} A_n J_0[j_n(x/l)^{\frac{1}{2}}].$$

This equation is not in a form such that the Fourier–Bessel theorem can be applied immediately but this can be remedied by writing $(x/l)^{\frac{1}{2}}$ as z and it is then seen that the A_n are the coefficients in the Fourier–Bessel expansion of $f(z^2l)$, $0 < z < 1$, i.e.

$$A_n = \frac{2}{J_1^2(j_n)} \int_0^1 zf(lz^2) J_0(j_n z) dz = \frac{2}{lJ_1^2(j_n)} \int_0^l f(x) J_0\left(\frac{j_n x^{\frac{1}{2}}}{l^{\frac{1}{2}}}\right) dx. \qquad \square$$

Problem 5.7 Solve $(l^2 - x^2)\partial^2 u/\partial x^2 - 2x\partial u/\partial x = \partial^2 u/\partial t^2$, $0 < x < l$, $t > 0$, under the conditions $u(0, t) = 0$, $u(l, t)$ bounded, $t > 0$; $u(x, 0) = x^3$, $\partial u(x, t)/\partial t = 0$, $t = 0$, $0 < x < l$. (This problem occurs when investigating the small transverse vibrations of a string rotating about a vertical axis.)

Solution. There will exist product solutions of the form $X(x)T(t)$ where

$$(l^2 - x^2)X'' - 2xX' + \lambda X = 0, \qquad T'' + \lambda T = 0.$$

The equation for X is Legendre's equation in the variable x/l but only positive values of the argument are considered. If X^* is defined by $X(x)$ for $x \geqslant 0$ and by $-X(-x)$ for $x \leqslant 0$ then X^* satisfies the same equation as X, is continuous for all x with $|x| \leqslant l$. X^* is also bounded at $x = \pm l$ and thus must be a Legendre polynomial of the variable x/l. Further, since X^* is odd in x it must be a Legendre polynomial of odd order, i.e. for $x > 0$ $X^* = X = P_{2n-1}(x/l)$ and $\lambda = 2n(2n-1)$, $n = 1, \dots$. The general form of solution is thus

$$u = \sum_{n=1}^{\infty} \{A_{2n-1} \cos[2n(2n-1)]^{\frac{1}{2}}t + B_{2n-1} \sin[2n(2n-1)]^{\frac{1}{2}}t\}P_{2n-1}(x/l),$$

the condition on $\partial u/\partial t$ gives $B_{2n-1} = 0$, and from the value of $u(x, 0)$ it is found that

$$x^3 = \sum_{n=1}^{\infty} A_{2n-1} P_{2n-1}(x/l).$$

The A_{2n-1} could be found by an extension of the analysis of Problem 4.4 but it is easier in this case to use the result $2P_3(x) = 5x^3 - 3x$, i.e. $x^3 = \frac{1}{5}[2P_3(x) + 3P_1(x)]$ and

$$x^3 = \frac{1}{5}l^3[2P_3(x/l) + 3P_1(x/l)],$$

i.e. $A_1 = 3l^3/5$, $A_3 = 2l^3/5$, $A_{2n-1} = 0$, $n \neq 1, 2$. $\qquad \square$

F

Problem 5.8 Obtain a solution of Laplace's equation within the sphere $r \leqslant a$, finite everywhere within this region, equal to unity on the sphere for $0 < \theta < \alpha$ and zero on the remainder of the sphere (r, θ being the usual spherical polar coordinates).

Solution. The appropriate form of solution to use is that of Problem 4.1, in this case however the finiteness condition within the sphere requires that $B_n = 0$ and the appropriate form of solution is $\displaystyle\sum_{n=0}^{\infty} A_n r^n P_n(\cos\theta)$ with

$$\sum_{n=0}^{\infty} A_n a^n P_n(\cos\theta) = \begin{cases} 1, & 0 < \theta < \alpha, \\ 0, & \alpha < \theta < \pi. \end{cases}$$

The A_n can now be obtained from Theorem 11 on writing x as $\cos\theta$; this gives $A_n a^n = \frac{1}{2}(2n+1)\int_{\cos\alpha}^{1} P_n(x)\,dx$. For $n = 0$ the integral is equal to $1 - \cos\alpha$ and for $n \neq 0$ it can be evaluated by use of the recurrence relation

$$(2n+1)P_n(x) = P'_{n+1}(x) - P'_{n-1}(x),$$

where the dash denotes the derivative with respect to the argument. Hence

$$A_n a^n = \frac{1}{2}[P_{n+1}(\cos\alpha) - P_{n-1}(\cos\alpha)], \qquad n \geqslant 1. \qquad \square$$

Problem 5.9 Solve

$$\frac{\partial^2 u}{\partial r^2} + \frac{1}{r}\frac{\partial u}{\partial r} = \frac{\partial u}{\partial t}, \qquad 0 < z < h, a < r < b,$$

under the conditions $u(a,t) = u(b,t) = 0, t > 0$; and $u(r,0) = 1, a < r < b$.

Solution. Separation of variables shows that there will be product solutions of the form $\{A\,J_0(\lambda r) + B\,Y_0(\lambda r)\}\exp(-\lambda^2 t)$ and the conditions on $r = a$ and $r = b$ give

$$A\,J_0(\lambda a) + B\,Y_0(\lambda a) = 0, \qquad A\,J_0(\lambda b) + B\,Y_0(\lambda b) = 0.$$

For non-zero A and B the determinant must vanish and thus

$$J_0(\lambda a)Y_0(\lambda b) - J_0(\lambda b)Y_0(\lambda a) = 0,$$

which is the equation satisfied by λ.

An equivalent method of calculating λ would be to use the fact that there will be product solutions of the form $R(r)\exp(-\lambda^2 t)$ where $R'' + R'/r + \lambda^2 R = 0$ and $R(a) = R(b) = 0$. The problem looked at from this point of view is of standard Sturm–Liouville type and from this general theory it is deduced that there will be an infinite number of values for λ (which we shall denote by δ_n, $n = 1, \ldots, \infty$) and the eigen functions will be orthogonal to each other with weight r. The nth eigen function is

$$J_0(\delta_n r)Y_0(\delta_n a) - Y_0(\delta_n r)J_0(\delta_n a)[\, = \phi_n].$$

The general form for u is thus

$$\sum_{n=1}^{\infty} a_n \exp(-\delta_n^2 t)[J_0(\delta_n r)Y_0(\delta_n a) - Y_0(\delta_n r)J_0(\delta_n a)],$$

where the a_n are determined by $\sum_{n=1}^{\infty} a_n \phi_n = 1$. On using the orthogonality property of ϕ_n, this gives

$$a_n \int_a^b r\phi_n^2 \, dr = \int_a^b r\,\phi_n(r) \, dr.$$

The evaluation of the integrals in the above expression is somewhat complicated, but it can be completed by using various relations involving Bessel functions [equations (1) and (10) of G. N. Watson, *Theory of Bessel Functions*, C.U.P., 1952, Chapter 5]. The final results are somewhat complicated and hence will not be quoted. □

5.2 Inhomogeneous Problems

The extension of the methods of the previous section to inhomogeneous problems will now be considered. The first step is to determine the set of functions ϕ_n which would be most appropriate to use, these can be found from considering the corresponding homogeneous problem as illustrated in the previous problems. The form of the solution is thus $\sum_{n=1}^{\infty} Y_n(y)\phi_n(x)$ and the next step is determining the equation for Y_n. The analysis will be restricted to equations of the form

$$\frac{1}{\rho(x)}\frac{\partial}{\partial x}\left(p(x)\frac{\partial u}{\partial x}\right) + a_3(x)u + b_1(y)\frac{\partial^2 u}{\partial y^2} + b_2(y)\frac{\partial u}{\partial y} + b_3(y)u = f,$$

on using the method of separation of variables it can be verified that the problem of solving the homogeneous problem reduces to a Sturm–Liouville problem and that $\rho(x)$ is the weight function. The above equation is then multiplied by $\rho\phi_n$ and integrated with respect to x from $x = 0$ to $x = c$. Integration by parts twice on the left hand side will give that this side of the equation will involve derivatives with respect to y of $\int_0^c \rho u\,\phi_n(x)\,[\text{i.e. } Y_n(y)\int_0^c \rho\phi_n^2 \, dx]dx$ and the prescribed boundary values. The right-hand side will be the known function $\int_0^c \rho\phi_n f(x,y)\,dx$; thus a differential equation will be obtained for the Y_n. For the case when the boundary conditions are homogeneous the equation for the Y_n can be obtained more simply by expanding f in terms of the appropriate ϕ_n. Substitution of this expansion and the corresponding expansion for u into the differential equation and equating coefficients of ϕ_n will then give the differential equation for Y_n.

77

In elliptic problems, when conditions are prescribed for two values of x and two values of y and one of the two sets of conditions is homogeneous then it is best to use an expansion which satisfies these conditions automatically.

It should again be remembered that the simplest method, if possible, is to reduce the inhomogeneous problem to a homogeneous one.

Problem 5.10 Solve

$$\frac{\partial^2 u}{\partial r^2} + \frac{1}{r}\frac{\partial u}{\partial r} - \frac{\partial^2 u}{\partial t^2} = -q \sin \omega t, \qquad 0 \leqslant r < a, \ t > 0,$$

under the conditions $u = \partial u/\partial t = 0$ at $t = 0$, $u = 0$, $r = a$; u finite for all r, $0 \leqslant r < a$, $a\omega$ not being a zero of $J_0(z)$.

Solution. The boundary conditions are homogeneous and it follows (on considering the case $q = 0$) by direct comparison with Problem 5.2 that the appropriate expansion is $u = \sum\limits_{n=1}^{\infty} J_0(j_n r/a)\psi_n(t)$ where we use the notation of Problem 5.2. The appropriate expansion for the right-hand side of the equation is obtained from Problem 4.7. Substitution in the equation and equating coefficients of $J_0(j_n r/a)$ gives

$$\psi_n'' + \frac{j_n^2}{a^2}\psi_n = \frac{2q \sin \omega t}{j_n J_1(j_n)}.$$

The conditions on u, $\partial u/\partial t$ at $t = 0$ show that $\psi_n = \psi_n' = 0$, $t = 0$, and, provided $j_n^2 \neq \omega^2 a^2$, the solution of the above equation satisfying these conditions is

$$\psi_n = \frac{2qa^2}{j_n^2 J_1(j_n)}\frac{[j_n \sin \omega t - \omega \sin j_n t]}{j_n^2 - \omega^2 a^2}. \qquad \square$$

Problem 5.11 Solve $\partial^2 u/\partial x^2 = \partial u/\partial t$, $0 < x < a$, $t > 0$, under the conditions $u(x, 0) = 1$, $0 < x < a$; $u(a, t) = 1$, $\partial u/\partial x = hu$, $x = 0$, $t > 0$, $h > 0$.

Solution. The simplest approach is to look for a solution which satisfies the conditions at $x = 0$ and $x = a$. Such a solution is $(1 + hx)/(1 + ha)$ and writing u as $(1 + hx)/(1 + ha) + v$ shows that v satisfies the same equation as u and v satisfies homogeneous conditions on $x = 0$ and $x = a$; this problem can then be solved by the previous methods.

We shall, however, consider the application of the direct technique described above. The homogeneous problem is that when the boundary condition on $x = a$ is $u = 0$, and by comparison with Problem 5.3 the

78

appropriate expansion for the homogeneous problem is

$$\sum_{n=1}^{\infty} X_n(x)\exp(-\lambda_n t) \text{ where } X'' + \lambda_n X = 0, \text{ and } X' = hX, x = 0; \quad X = 0,$$
$x = a$.

The solution for X_n satisfying the condition on $x = a$ is $B_n \sin \lambda_n^{\frac{1}{2}}(x-a)$ and the condition on $x = a$ gives $\lambda^{\frac{1}{2}}\cos \lambda^{\frac{1}{2}}a = -h \sin \lambda^{\frac{1}{2}}a$. Thus $\lambda^{\frac{1}{2}}a = w_n$, where w_n are the roots of $ha \tan w = -w$. The form of solution to be sought is thus $u = \sum_{n=1}^{\infty} \psi_n(t)\sin w_n(x-a)/a$ and

$$\psi_n \int_0^a \sin^2 w_n(x-a)/a \, dx = \int_0^a u(x,t)\sin w_n(x-a)/a \, dx.$$

It can be shown by integration by parts twice and using the boundary condition on $x = 0, a$ and the definition of w_n that

$$\int_0^a \frac{\partial^2 u}{\partial x^2} \sin \frac{w_n(x-a)}{a} \, dx = -\frac{w_n}{a} - \frac{w_n^2}{a^2}\int_0^a u \sin \frac{w_n(x-a)}{a} \, dx.$$

Hence multiplying both sides of the partial differential equation by $\sin w_n(x-a)/a$ and integrating from $x = 0$ to $x = a$ gives, on using the definition of ψ_n in terms of u,

$$\frac{d\psi_n}{dt} = -\frac{w_n^2}{a^2}\psi - w_n \Big/ a \int_0^a \sin^2 w_n(x-a)/a \, dx$$

$$= -\frac{w_n^2}{a^2}\psi_n - \frac{4w_n^2}{a^2(2w_n - \sin 2w_n)}.$$

Thus
$$\psi_n = A_n \exp\left(-\frac{w_n^2 t}{a^2}\right) - \frac{4}{(2w_n - \sin 2w_n)}.$$

The condition on $t = 0$ gives $1 = \sum_{n=1}^{\infty} \psi_n(0)\sin w_n(x-a)/a$, i.e.

$$\int_0^a \sin \frac{w_n(x-a)}{a} \, dx = \psi_n(0) \int_0^a \sin^2 \frac{w_n(x-a)}{a} dx$$

or
$$\frac{a}{w_n}(\cos w_n - 1) = \frac{a\psi_n(0)(2w_n - \sin 2w_n)}{4w_n}.$$

Hence
$$A_n = \frac{4\cos w_n}{(2w_n - \sin 2w_n)}. \qquad \Box$$

Problem 5.12 Solve

$$\frac{\partial^2 u}{\partial r^2} + \frac{1}{r}\frac{\partial u}{\partial r} + \frac{\partial^2 u}{\partial z^2} = 0, \qquad 0 < r < a, \quad 0 < z < h,$$

under the conditions

$u(0, z)$ finite, $u(a, z) = u_1$, $0 < z < h$; $u(r, 0) = u_2$, $u(r, h) = u_3$ where u_1, u_2, u_3 are constants.

Solution. The boundary conditions on $z = 0$ and $z = h$ are simple so that it is worth considering the possibility of obtaining homogeneous conditions at $z = 0$ and $z = h$. The function $u_2(1 - z/h) + u_3 z/h$ satisfies the conditions at $z = 0$ and $z = h$, and if u is written as $u_2(1 - z/h) + u_3 z/h + v$ then v will satisfy the same equation as u; v vanishes at $z = 0$ and $z = h$ and is equal to $u_1 - u_3 z/h - u_2(1 - z/h)$ on $r = a$.

This problem for v is of the standard type encountered in the previous chapter and can be solved by assuming a Fourier sine series representation.

An alternative method of solution would be to attempt to obtain a series representation for u which would automatically satisfy the condition on $r = a$. One such solution is u_1; writing u as $w + u_1$ shows that w satisfies the same equation as u, vanishes on $r = a$ and is equal to $u_2 - u_1$ and $u_3 - u_1$ on $z = 0$ and $z = h$ respectively. The problem for w can be solved by use of separation of variables and by analogy with previous examples it can be found that the appropriate expansion is the Fourier–Bessel series. $\qquad\qquad\square$

In examples of this type where two approaches are possible, it is preferable to use the one with least complications. In particular, if one set of conditions are homogeneous one should use the approach which automatically satisfies these conditions. Thus if u_2 and u_3 had been zero the sine series approach would be the less complicated, and for $u_1 = 0$ use of the Fourier–Bessel method would have advantages. There is the additional point that integrals involving Bessel functions are usually not as simple as those involving trigonometric functions, and this could mean that even for non-homogeneous conditions it might be better to use trigonometric series where possible.

EXERCISES

1. Solve

$$\frac{\partial^2 u}{\partial r^2} + \frac{1}{r}\frac{\partial u}{\partial r} + \frac{\partial^2 u}{\partial z^2} = 0, \qquad 0 \leqslant r < a, z > 0,$$

subject to the conditions $u(0, z)$ finite, $u(a, z) = 0$, $z > 0$, $u(r, 0) = 1$, $0 \leqslant r < a, u \to 0$ as $z \to \infty$.

2. Obtain a solution of Laplace's equation finite everywhere within the

sphere $r = a$ and equal to $\cos^3\theta$ on the boundary of the sphere, r, θ being the usual polar coordinates.

Obtain also the solution of Laplace's equation valid outside the sphere, tending to zero at infinity and taking the same values on the boundary.

3.   Solve

$$\frac{\partial^2 u}{\partial r^2} + \frac{1}{r}\frac{\partial u}{\partial r} = \frac{\partial^2 u}{\partial t^2}, \qquad 0 \leqslant r < a, t > 0,$$

subject to the conditions $\partial u/\partial r = 0, r = a, u(0, t)$ finite, $t > 0; u(r, 0) = r$, $\partial u/\partial t = r^3, t = 0, 0 \leqslant r < a$.

4.   Solve $\partial u/\partial t = \partial^2 u/\partial x^2$, $0 < x < l, t > 0$, subject to the conditions $\partial u/\partial x = hu, x = 0, \partial u/\partial x = -hu, x = l, t > 0; u(x, 0) = 1, 0 < x < l$, h being a constant.

5.   Solve the boundary value problem of the previous exercise when the conditions on $x = 0, x = l$ are $\partial u/\partial x = hu - hu_1, \partial u/\partial x = -hu + hu_1$ respectively where u_1 is a constant.

Answers to Exercises

Chapter 1

1. (iv) $\pi^2/8$, $\pi^4/96$.

2. $x = 2\sum_{n=1}^{\infty} \frac{(-1)^{n+1}\sin nx}{n}$, $\quad 0 \leqslant x < \pi$;

$\cos\frac{1}{2}x = \frac{2}{\pi} - \frac{4}{\pi}\sum_{n=1}^{\infty}\frac{(-1)^{n+1}\cos nx}{1-4n^2}$, $\quad 0 \leqslant x \leqslant c$.

4. $4\sum_{n=1}^{\infty}\frac{[(-1)^{n+1}(2n-1)\pi\exp(c)-2c]}{(2n-1)^2\pi^2+4c^2}\frac{\cos(2n-1)\pi x}{2c}$,

$4\sum_{n=1}^{\infty}\frac{[2(-1)^{n+1}c\exp c+(2n-1)\pi]}{[(2n-1)^2\pi^2+4c^2]}\frac{\sin(2n-1)\pi x}{2c}$.

5. $\frac{2}{\pi} - \frac{4}{\pi}\sum_{n=1}^{\infty}\frac{\cos 2nx}{4n^2-1}$.

Chapter 2

1. $\frac{4qa}{\pi^2}\sum_{n=1}^{\infty}\frac{\sinh[(2n-1)\pi y/a]\text{sech}[(2n-1)\pi b/a]\sin[(2n-1)\pi x/a]}{(2n-1)^2}$.

2. $\rho\sin\theta, \frac{1}{2}\rho\sin\theta + \frac{1}{\pi} - \frac{2}{\pi}\sum_{n=1}^{\infty}\frac{\rho^{2n}\cos 2n\theta}{4n^2-1}$ (see Problem 1.6).

3. $\frac{4}{\pi}\sum_{n=1}^{\infty}\frac{\sinh[(2n-1)\pi y/2a]\text{cosech}[(2n-1)\pi b/2a]\sin[(2n-1)\pi x/2a]}{(2n-1)}$.

4. $u-1$ is a solution of exercise 1.

5. $a+(b-a)\frac{x}{l} - \frac{4a}{\pi}\sum_{n=1}^{\infty}\sin[(2n-1)\pi x/l]\exp[-(2n-1)^2\pi^2 t/l^2](2n-1)^{-1}$

$\qquad -\frac{21}{\pi}(b-a)\sum_{n=1}^{\infty}(-1)^{n+1}\sin(n\pi x/l)\exp[-n^2\pi^2 t/l^2]n^{-1}$.

6. $x - \frac{8l}{\pi^2}\sum_{n=1}^{\infty}(-1)^{n+1}\frac{\cos[(2n-1)\pi t/2l]\sin[(2n-1)\pi x/2l]}{(2n-1)^2}$.

7. $2\sum_{n=1}^{\infty}\frac{[n\pi\sin t - l\sin n\pi t/l]}{(n^2\pi^2-l^2)}\frac{\sin n\pi x}{l}$.

Chapter 3

2. $2\beta/\beta^4 + 4$.

3. $2(\cos x - \cos 2x)/\pi x$.

4. $\dfrac{2}{\pi}\displaystyle\int_0^\infty \dfrac{\sin \beta x}{\beta^4 + \omega^2}\left[\beta^3\sin \omega t - \beta\omega \cos \omega t + \beta\omega \exp(-\beta^2 t)\right] d\beta$.

   (The first two terms can be evaluated explicitly using Tables of Fourier transforms (Erdelyi, A., *et al. Tables of Integral Transforms*, McGraw-Hill, 1954, vol. 1, p. 66, nos. (23) and (24)) and are equal to $\exp[-x(\omega/2)^{\frac{1}{2}}]\sin[\omega t - x(\omega/2)^{\frac{1}{2}}]$.)

5. $\dfrac{2}{\pi}\displaystyle\int_0^\infty \dfrac{\cos \beta x}{\beta^2 + 2}[\exp(-\beta^2 t) - \exp(-t)] \, d\beta$.

Chapter 4

1. $\dfrac{8}{63}P_5 + \dfrac{4}{9}P_3 + \dfrac{3}{7}P_1$.

2. See Problem 5.8.

3. $2\displaystyle\sum_{n=1}^\infty \dfrac{J_0(j_n x)}{j_n^2 J_1^2(j_n)}[j_n J_1(j_n) - 2J_2(j_n)]$, using notation of Problem 5.2.

4. $(2 + 3\lambda)\sin \lambda^{\frac{1}{2}}\pi + \lambda^{\frac{1}{2}}\cos \lambda^{\frac{1}{2}}\pi = 0$.

Chapter 5

1. (The notation of Problem 5.2 is used in 1 and 3.)

   $2\displaystyle\sum_{n=1}^\infty \dfrac{J_0(j_n r/a)}{j_n J_1(j_n)} \exp(-j_n z/a)$.

2. $\frac{2}{5}r^3 P_3(\cos \theta) + \frac{3}{5}r \cos \theta, \qquad \frac{2}{5}r^{-4}P_3(\cos \theta) + \frac{3}{5}r^2\cos \theta$.

3. $2a\displaystyle\sum_{n=1}^\infty J_0(j_n r/a)[j_n^2 J_1(d_n)\cos j_n t/a + a^3[j_n J_1(d_n)$
   $\qquad\qquad\qquad - 2J_2(j_n)]\sin d_n t/a]/d_n^3 J_1^2(d_n)$.

4. $2\displaystyle\sum_{n=1}^\infty \dfrac{[\lambda_n \sin \lambda_n l + h - h \cos \lambda_n l][\cos \lambda_n x + h\lambda_n^{-1}\sin \lambda_n x]\exp(-\lambda_n^2 t)}{(\lambda_n^2 + h^2) + 2h}$.

   where λ_n are the roots of $(\lambda^2 - h^2)\tan \lambda l = 2h\lambda$.

5. Writing u as $u_1 + v$ gives that v is $(1 - u_1)$ times the solution of exercise 4.

Appendix 1. Some Simple Fourier Series

1. $\displaystyle\sum_{n=1}^{\infty}\frac{\sin(n\pi x/c)}{n}=\frac{\pi}{2}\frac{(1-x)}{c},\quad 0<x\leqslant c,$

2. $\displaystyle\sum_{n=1}^{\infty}\frac{\cos(n\pi x/c)}{n}=\frac{-\ln(2\sin\pi x)}{2c},\quad 0<x\leqslant c,$

3. $\displaystyle\sum_{n=1}^{\infty}\frac{(-1)^{n-1}\sin n\pi x/c}{n}=\frac{\pi x}{2c},\quad 0\leqslant x<c,$

4. $\displaystyle\sum_{n=1}^{\infty}\frac{(-1)^{n-1}\cos n\pi x/c}{n}=\ln\left(2\cos\frac{\pi x}{2c}\right),\quad 0\leqslant x<c,$

5. $\displaystyle\sum_{n=1}^{\infty}\frac{\cos n\pi x/c}{n^2}=\pi^2\left(\frac{1}{6}-\frac{x}{2c}+\frac{x^2}{4c^2}\right),\quad 0\leqslant x\leqslant c,$

6. $\displaystyle\sum_{n=1}^{\infty}\frac{(-1)^{n-1}\cos n\pi x/c}{n^2}=\pi^2\left(\frac{1}{12}-\frac{x^2}{4c^2}\right),\quad 0\leqslant x\leqslant c,$

7. $\displaystyle\sum_{n=1}^{\infty}\frac{\sin n\pi x/c}{n^3}=\pi^3\left(\frac{x}{6c}-\frac{x^2}{4c^2}+\frac{x^3}{12c^3}\right),\quad 0\leqslant x\leqslant c,$

8. $\displaystyle\sum_{n=1}^{\infty}\frac{(-1)^{n-1}\sin n\pi x/c}{n^3}=\frac{\pi^3 x}{12c}\left(1-\frac{x^2}{c^2}\right),\quad 0\leqslant x\leqslant c.$

Appendix 2.   Table of Fourier Integrals

	$f(x)$	$F(\beta) = \int_{-\infty}^{\infty} \exp(-i\beta x) f(x)\, dx$
1.	$\lambda^{-1} f(x\lambda^{-1})$	$F(\lambda\beta)$
2.	$F(x)$	$2\pi f(-\beta)$
3.	$f(\lambda x)\exp ibx$	$\lambda^{-1} F[(\beta - b)/\lambda]$
4.	$(a^2 + x^2)^{-1}$	$(\pi/a)\exp - a|\beta|,\ a > 0$
5.	$\exp - x^2$	$\pi^{\frac{1}{2}}\exp - \tfrac{1}{4}\beta^2$

	$f(x)$	$F_s(\beta) = \int_0^{\infty} f(x)\sin\beta x\, dx$
6.	$F_s(x)$	$\tfrac{1}{2}\pi f(\beta)$
7.	$\lambda^{-1} f(x\lambda^{-1})$	$F_s(\lambda\beta)$
8.	$x^{-1}\sinh\alpha x\operatorname{cosech}\gamma x$	$\tan^{-1}[\tan(\tfrac{1}{2}\pi\alpha\gamma^{-1})\tanh(\tfrac{1}{2}\pi\gamma^{-1}\beta)]$
9.	$x^{-1}\exp - ax,\ a > 0$	$\tan^{-1} a^{-1}\beta$
10.	$\exp - ax$	$\beta(\alpha^2 + \beta^2)^{-1}$

	$f(x)$	$F_c(\beta) = \int_0^{\infty} f(x)\cos\beta x\, dx$
11.	$F_c(x)$	$\tfrac{1}{2}\pi\, f(\beta)$
12.	$\exp - \alpha x^2$	$\tfrac{1}{2}\pi^{\frac{1}{2}}\alpha^{-\frac{1}{2}}\exp - \beta^2/4\alpha$
13.	$\cosh\alpha x\operatorname{sech}\gamma x$	$\pi\gamma^{-1}\dfrac{\cos(\tfrac{1}{2}\pi\gamma^{-1}\alpha)\cosh\tfrac{1}{2}\pi\gamma^{-1}\beta}{\cos\pi\alpha\gamma^{-1} + \cosh\pi\gamma^{-1}\beta}$

Index